Surfactants - Fundamental Concepts and Emerging Perspectives

Edited by Olasehinde Owoseni

Published in London, United Kingdom

Surfactants - Fundamental Concepts and Emerging Perspectives
http://dx.doi.org/10.5772/intechopen.105270
Edited by Olasehinde Owoseni

Contributors
Harasit Kumar Mandal, Ali Khalfallah, Mona Kharazi, Javad Saien, Safa Gamoudi, Ezzeddine Srasra, Chidi Obi, Mary-Ann N. Mgbemena, Olasehinde Owoseni

First published in London, United Kingdom, 2024 by IntechOpen
IntechOpen is the global imprint of INTECHOPEN LIMITED, registered in England and Wales, registration number: 11086078, 5 Princes Gate Court, London, SW7 2QJ, United Kingdom

British Library Cataloguing-in-Publication Data
A catalogue record for this book is available from the British Library

Additional hard and PDF copies can be obtained from orders@intechopen.com

Surfactants - Fundamental Concepts and Emerging Perspectives
Edited by Olasehinde Owoseni
p. cm.
Print ISBN 978-1-83768-706-0
Online ISBN 978-1-83768-707-7
eBook (PDF) ISBN 978-1-83768-708-4

For EU product safety concerns:
IN TECH d.o.o., Prolaz Marije Krucifikse Kozulić 3, 51000 Rijeka, Croatia,
info@intechopen.com or visit our website at intechopen.com.

Meet the editor

Olasehinde Owoseni earned his BS in Chemical Engineering from Obafemi Awolowo University, Nigeria, and a Ph.D. in Chemical and Biomolecular Engineering from Tulane University, USA. His doctoral research was on the colloidal and interfacial phenomena of silica-based nanomaterials, surfactants, and amphiphilic biopolymers to design more effective and benign oil spill remediation technology. Dr. Owoseni joined Intel Corporation in 2016 as a process technology development engineer making contributions to semiconductor device manufacturing technology. He transitioned to the equipment development engineering role in 2020, enabling atomic layer deposition (ALD) process equipment development and supply chain solutions for ultra-thin film deposition in semiconductor chip fabrication. Outside work, he enjoys playing soccer, hiking, playing the keyboard, and spending time with his family.

Contents

Preface

Surfactants are amphiphilic molecules that are used across a wide range of products and applications, including consumer products, medicines, remediation agents, and agrochemical products. This book presents an overview of the fundamentals of surfactant systems and links the fundamentals to emerging research and technological directions. This book is an excellent resource for new and experienced professionals across academia and industry.

The first three chapters in the book provide a comprehensive foundation on surfactants. Chapter 1 is the introductory chapter. Chapter 2 captures the key fundamentals of surfactants. This is an excellent resource for students and professionals who are new to the field. Chapter 3 focuses on the emerging opportunities with environmentally benign surfactants. Environmental aspects are now a critical criterion in the design and application of surfactants.

This book then provides detailed deep dives on three key topics in Chapters 4–6. Chapter 4 is on organoclays formed by combining surfactants with naturally occurring clay minerals. The interactions of surfactant and clay minerals are interesting from a research standpoint but have been demonstrated to have superior characteristics in practical applications such as oil spill remediation. Chapter 5 provides the properties and application of ionic liquid-based surfactants in chemically enhanced oil recovery. The content in Chapters 4 and 5 mostly involves the use of experimental techniques in characterizing surfactant systems. However, the combination of experiments with modeling techniques provides additional insights that are useful in understanding complex surfactant systems. Chapter 6 focuses on the models developed for surfactant characterization in complex microheterogenous systems. Overall, this book bridges the fundamental concepts of surfactant systems in Chapters 2 and 3 and emerging perspectives in Chapters 4 and 5.

I would like to thank all the authors who contributed to this book, including Ali Khalfallah, Chidi Obi, Mary-Ann N. Mgbemena, Safa Gamoudi, Ezzeddine Srasra, Mona Kharazi, Javad Saien, and Harasit Kumar Mandal. I would also like to thank Publishing Process Manager Ana Javor at IntechOpen for her tremendous support. I would also like to thank the Commissioning Editor Marijana Josipovic at IntechOpen for the excellent support and guidance throughout the publication process. Thank you all.

Olasehinde Owoseni
Intel Corporation,
Hillsboro, OR, USA

Chapter 1

Introductory Chapter: Surfactants – Bridging Fundamental Concepts with Emerging Perspectives

Olasehinde Owoseni

1. Introduction

Surfactants are amphiphilic molecules with hydrophilic head and hydrophobic tail groups (**Figure 1**). The amphiphilicity of these surfactant molecules drives their assembly into a wide range of structures including micelles, reverse micelles, liquid crystalline mesophases, and vesicles [1]. These surfactant assemblies have been exploited as templates for nanostructured-materials synthesis [2, 3]. Surfactants are applied in a wide range of applications including the formulation of detergents, medicines, vaccines, paints, cosmetics, oil spill dispersants, enhanced oil recovery agents, and agrochemicals products. This book provides an overview of the fundamentals, emerging perspectives, and applications of surfactants.

Typical considerations in the study and application of surfactant systems are built on their fundamental assembly and physicochemical properties (**Figure 2**). For effectiveness in technological applications, surfactant is typically applied as mixed systems including mixtures of multiple surfactants, particles, and polymers [1, 4–9]. The novelty of research efforts on surfactant can include synergism with nanomaterials, process scalability and environmental impacts (**Figure 2**).

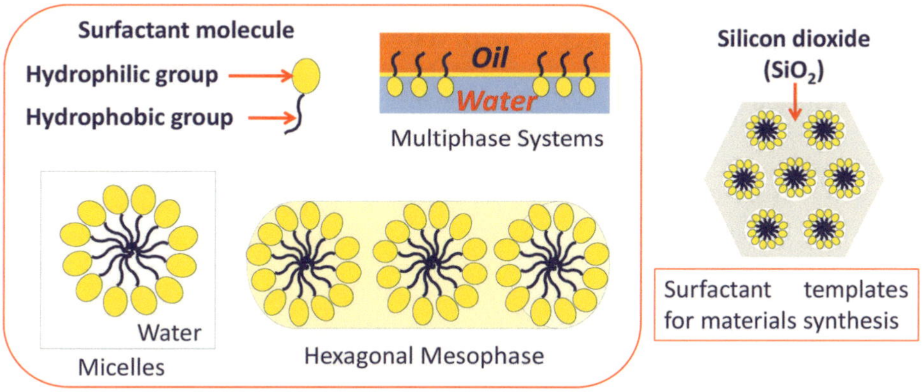

Figure 1.
Surfactant self-assembly and microstructures.

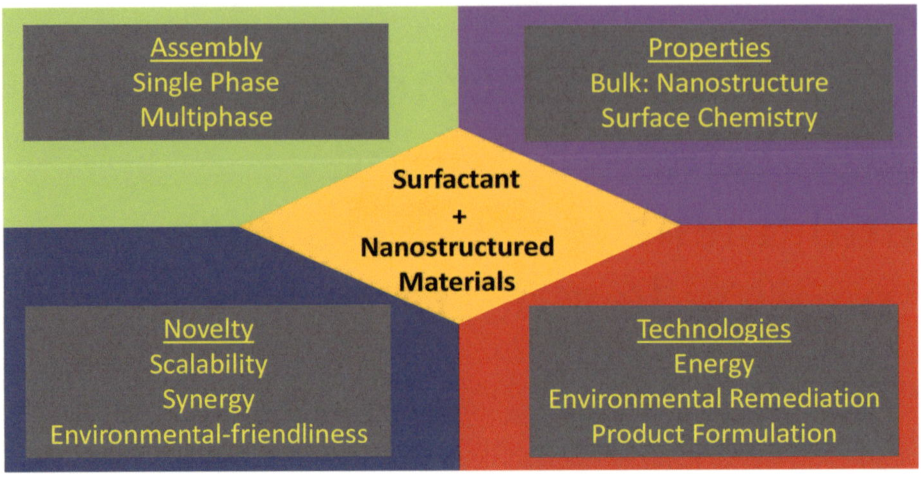

Figure 2.
General considerations in the study and application of surfactant systems.

The first two chapters of this book present a fundamental review of surfactant systems. These introductory chapters should be an excellent introduction to the fundamental phenomena of surfactants. Chapter 1 covers a broad scope of the structure and application of surfactants. However, concerns about the potential impacts of surfactant in our everyday lives has stimulated recent research efforts in the development and application of environmentally benign surfactant systems [10–14]. Thus, Chapter 2 provides a more-tailored perspective on the development and application of eco-benign surfactants.

Chapter 3 presents an interesting topic of combining surfactants with clay minerals. The combination of surfactant and clay minerals such as halloysite clay nanotubes has attracted significant attention in environmental remediation applications [15, 16]. The unique hollow geometry of the interfacially adherent particles allows for surfactant encapsulation and delivery to the oil-water interface. Surfactant release lowers the interfacial tension and facilitates oil dispersion [15]. In general, the synergistic combination of surfactant and clay minerals in organoclays can provide significant benefits in several practical applications beyond oil spill remediation. The characteristics of the organoclays formed from the clay-surfactant interactions are determined by the individual characteristics of the surfactant and clay mineral as well as the method of preparation. Thus, the chapter starts with a fundamental discussion on surfactants including their classification, physicochemical, and interfacial adsorption characteristics. This is followed by a section focused on clay minerals. The classification of clay minerals, chemical structure, and fundamental properties such as swelling capacity, cation exchange capacity, surface area, and charge are presented. The interaction of surfactant and clay minerals is then discussed with critical details on the preparation methods and prevailing physicochemical mechanisms. The chapter also provides insights into the structure of the resulting surfactant-modified clay minerals (organoclays) and their practical applications.

The use of Ionic Liquid-Based Surfactants in Chemical Enhanced Oil Recovery (CEOR) is presented in Chapter 4. Surfactants play a critical role in reducing the Interfacial tension (IFT) [15] and altering surface wettability [17]. However, the tunability and stability of ionic liquids in harsh environmental conditions make them attractive for CEOR [18, 19]. The fundamental characteristics of ionic-liquid-based surfactants and their relevance to Chemical Enhanced Oil Recovery (CEOR) mechanisms are first discussed in the chapter. The discussion includes the critical physical properties and chemical structure of Ionic Liquid-based surfactants. The chapter briefly introduces the various enhanced oil recovery methods including gas injection, thermal methods, chemical methods, and new emerging methods in the field. The next section in the chapter then provides a deep dive into the chemical methods spanning surfactant flooding, micellar flooding, and alteration of reservoir rock wettability from oil-wet to water-wet. The efficacy of surfactant systems in CEOR is strongly dependent on IFT reduction. The last section discusses the effectiveness of Ionic Liquid-Based Surfactants relative to conventional surfactants in CEOR. These are key considerations that are relevant in surfactant screening for CEOR applications [20].

Surfactants have found application in microheterogeneous catalysis with superior properties for catalytic reactions and process tunability relative to homogeneous reactions [21, 22]. This is especially relevant in reactions where all the reactants cannot be solubilized in a single solvent [21]. These reaction systems are complex with the presence of reactants; multiple to effectively understand the surfactant mechanisms in complex microheterogenous catalysis applications, combining experimental and modeling techniques is beneficial [23, 24]. The last chapter in this book provides a unique perspective on the models that have been developed to explain the surfactant effects on reaction rates. Overall, this book bridges the fundamental concepts and emerging applications of surfactant systems.

Author details

Olasehinde Owoseni
Intel Corporation, United States of America

*Address all correspondence to: sehindeowoseni@yahoo.com

References

[1] Olasehinde Owoseni YZ. Microstructural characteristics of surfactant assembly into a gel-like mesophase for application as an oil spill dispersant. Journal of Colloid and Interface Science. 2018;**524**:279-288

[2] Holmberg K. Surfactant-templated nanomaterials synthesis. Journal of Colloid and Interface Science. 15 Jun 2004;**274**(2):355-364

[3] Su Y, Wang Y, Owoseni O, Zhang Y, Gamliel DP, Valla JA, et al. A bottle-around-a-ship method to generate hollow thin-shelled particles containing encapsulated iron species with application to the environmental decontamination of chlorinated compounds. ACS Applied Materials & Interfaces. 25 Apr 2018;**10**(16):13542-13551. DOI: 10.1021/acsami.7b14308. Epub 2018 Apr 12. PMID: 29620856

[4] Maestro A, Santini E, Zabiegaj D, Llamas S, Ravera F, Liggieri L, et al. Particle and particle-surfactant mixtures at fluid interfaces: Assembly, morphology, and rheological description. Advances in Condensed Matter Physics. 2015;**2015**:17. Article ID 917516. DOI: 10.1155/2015/917516

[5] Francesca Ravera MF. Liquid–liquid interfacial properties of mixed nanoparticle–surfactant systems. Colloids and Surfaces A: Physicochemical and Engineering Aspects. 2008;**323**(1-3):99-108

[6] Olasehinde Owoseni YS. Hydrophobically modified chitosan biopolymer connects halloysite nanotubes at the oil-water interface as complementary pair for stabilizing oil droplets. Journal of Colloid and Interface Science. 2022;**620**:135-143

[7] Olasehinde Owoseni YZ. Tuning the wettability of halloysite clay nanotubes by surface carbonization for optimal emulsion stabilization. Langmuir. 2015;**31**(51):13700-13707

[8] Pal JY. Investigation of surfactant-polymer interactions using rheology and surface tension measurements. Polymers. 2020;**12**(10):2302

[9] Ritacco HA. Polyelectrolyte/surfactant mixtures: A pathway to smart foams. ACS Omega. 2022;7(41):36117-36136

[10] Bhadani A, Kafle A, Ogura T, Akamatsu M, Sakai K, Sakai H, et al. Current perspective of sustainable surfactants based on renewable building blocks. Current Opinion in Colloid & Interface Science. 2020;**45**:124-135. DOI: 10.1016/j. cocis.2020.01.002. ISSN 1359-0294

[11] Farias CBB, Almeida FCG, Silva IA, Souza TC, Meira HM, Soares da Silva RCF, et al. Production of green surfactants: Market prospects. Electronic Journal of Biotechnology. 2021;**51**:28-39

[12] Giwa A, Chalermthai B, Shaikh B, Taher H. Green dispersants for oil spill response: A comprehensive review of recent advances. Marine Pollution Bulletin. Aug 2023;**193**:115118

[13] John V, Arnosti C, Field J, Kujawinski EB, MacCormick A. The Role of Dispersants in Oil Spill Remediation: Fundamental Concepts, Rationale for Use, Fate, and Transport Issues. 2016. DOI: 10.17615/epc8-d024

[14] Athas JC, Jun K, McCafferty C, Owoseni O, John VT, Raghavan SR. An effective dispersant for oil spills based on food-grade amphiphiles. Langmuir. 12 Aug 2014;**30**(31):9285-9294.

DOI: 10.1021/la502312n. Epub 2014 Jul 29. PMID: 25072867

[15] Olasehinde Owoseni EN. Release of surfactant cargo from Interfacially-active Halloysite clay nanotubes for oil spill remediation. Langmuir. 2014;**30**(45):13533-13541

[16] Owoseni O, Nyankson E, Zhang Y, Adams DJ, He J, Spinu L, et al. Interfacial adsorption and surfactant release characteristics of magnetically functionalized halloysite nanotubes for responsive emulsions. Journal of Colloid and Interface Science. 1 Feb 2016;**463**:288-298. DOI: 10.1016/j.jcis.2015.10.064. Epub 2015 Oct 30. PMID: 26555959

[17] Achinta Bera HB. Ionic liquids as alternatives of surfactants in enhanced oil recovery—A state-of-the-art review. Journal of Molecular Liquids. 2016;**224**(Part A):177-188

[18] Somoza A, Arce A, Soto A. Oil recovery tests with ionic liquids: A review and evaluation of 1-decyl-3-methylimidazolium triflate. Petroleum Science. 2022;**19**(4):1877-1887. DOI: 10.1016/j.petsci.2021.10.025. ISSN 1995-8226

[19] Nandwani S, Malek N, Chakraborty M, Gupta S. Insight into the application of surface-active ionic liquids in surfactant based enhanced oil recovery processes–a guide leading to research advances. Energy & Fuels. 2020. DOI: 10.1021/acs.energyfuels.0c00343

[20] Benzagouta MS, AlNashef IM, Karnanda W, Al-Khidir K. Ionic liquids as novel surfactants for potential use in enhanced oil recovery. Korean Journal of Chemical Engineering. 2013;**30**(11):2108-2117

[21] Schwarze M, Pogrzeba T, Volovych I, Schomäcker R. Microemulsion systems for catalytic reactions and processes. Catalysis Science & Technology. 2015;**5**:24-33

[22] Lin J-M, Yamada M. Microheterogeneous systems of micelles and microemulsions as reaction media in chemiluminescent analysis. TrAC Trends in Analytical Chemistry. 2003;**22**(2):99-107. ISSN 0165-9936

[23] Bravo-Díaz C, Romsted LS, Liu C, Losada-Barreiro S, Pastoriza-Gallego MJ, Gao X, et al. To model chemical reactivity in heterogeneous emulsions, think homogeneous microemulsions. Langmuir. 25 Aug 2015;**31**(33):8961-8979. DOI: 10.1021/acs.langmuir.5b00112. Epub 2015 Apr 16. PMID: 25805058

[24] Kovalchuk VI, EV. Surfactant adsorption layers: Experiments and modeling. Langmuir. 2023;**39**(10):3537-3545

Chapter 2

Structure and Applications of Surfactants

Ali Khalfallah

Abstract

Surfactant molecules have two parts, a lipophilic (apolar) part that retains fat and a hydrophilic (polar) part that is miscible with water. The lipophilic portion consists of one or more aliphatic, straight or branched or aromatic or even alkylaromatic hydro- or fluorocarbon chains. The hydrophilic portion or polar head consists of one or more polar groups, ionic or nonionic. Surfactants have a wide variety of applications that include membrane permeabilization and dissolution, inclusion body solubilization, as well as membrane protein solubilization, biochemistry, crystallization, and manipulation. The behavior of these molecules is directly related to the aversion to water of the nonpolar groups, whereas the polar moieties tend to be highly hydrated. Their surfactant properties are therefore essentially based on the balance between the hydrophilic and hydrophobic parts of the molecule, called HLB (Hydrophile-Lipophile Balance).

Keywords: surfactanst, structures, applications, monotailed, bitailed, bolaphiles, hydro-fluorocarbon, chains

1. Introduction

1.1 General

Surfactant molecules are amphiphilic compounds. These molecules have two parts with different affinities with water. One part is polar and soluble in water, commonly called "hydrophilic head", while the other part is apolar, presenting rather an affinity for fatty substances and commonly called "hydrophobic chain." This feature gives surfactant molecules the property of adsorbing preferentially at interfaces and thus reducing the interfacial tension to improve the dispersion or solubilization of one phase in another immiscible one.

1.2 Properties of surfactants

The amphiphilic nature of surfactant molecules induces a very strong tendency to migrate at the interfaces [1], which allows them to improve the wettability of a liquid, to stabilize a foam or an emulsion, to promote the dispersion of a solid in a liquid [2, 3].

IntechOpen

The physico-chemical properties of surfactants, which determine their potential applications, depend on several physical quantities [4].

1.2.1 Hydrophilic/lipophilic balance

To determine the hydrophilic power of a surfactant, the HLB (Hydrophilic Lipophilic Balance) defined by Griffin [5] can be used. The HLB of a surfactant is related to its solubility. It extends on a scale from 0 to 20, and the higher the value, the more the surfactant is soluble in water (more hydrophilic, **Table 1**).

The HLB value of surfactants can be calculated by the empirical method of Griffin (Eq. 1) [5], with H and L the respective molar masses of the hydrophilic part and the lipophilic part. However, this relationship is not suitable for other surfactants, since it does not take into account counterions in the case of ionic surfactants, nor the existence of establishments or functional groups on the hydrophobic chain.

$$\mathrm{HLB} = \left(\frac{H}{H+L}\right) + 20 \tag{1}$$

1.2.2 Surface tension

Surface tension is the tension that exists at the interface between a gas and a liquid. This tension accounts for the force exerted by the molecules which are on the surface separating two media. In the presence of surfactants on the surface, this force decreases.

1.2.3 Critical micellar concentration

The Critical micellar concentration (CMC) is the concentration above which the interface is saturated and part of the surfactant molecules come together in the form of aggregates called micelles. Beyond that, the surface tension is no longer modified, even at concentrations clearly above the CMC

1.3 Classification of surfactants

There are different possible classifications of surfactants. They can be classified according to:

Surfactant properties	HLB value	Solubility
Defoamer	1.5–3	↑ Lipophilic
Water-in-oil emulsifier	3–6	
Wetting	7–9	
Oil in water emulsifier	8–18	Hydrophilic ↓
Detergent	13–15	
Peptizer-solubilizer	15–18	

Table 1.
Classification of surfactants according to value by HLB [6].

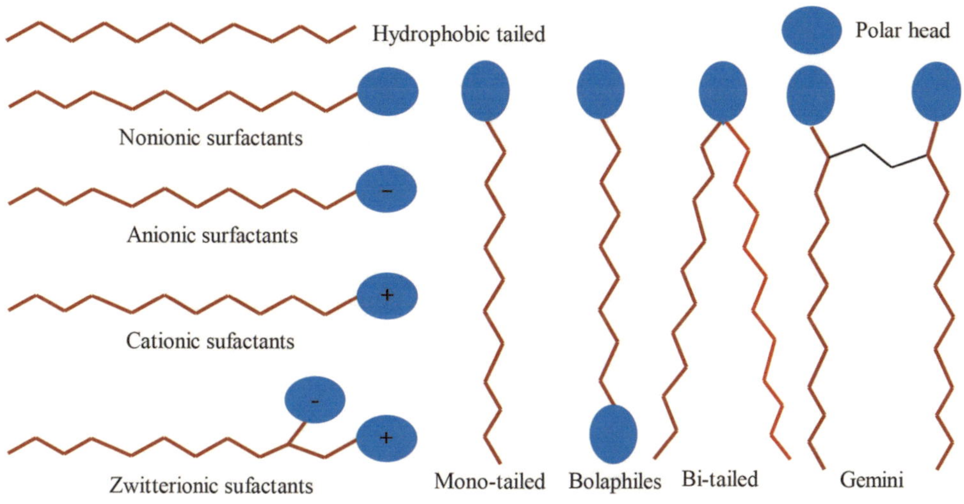

Figure 1.
Schematic representations of surfactants according to their polar heads and hydophobic chain.

- the nature of their polar head (nonionic, anionic, cationic or amphoteric) (**Figure 1**).
- number of polar head and hydrophobic chain which makes it possible to classify surfactants as mono-chain, bi-chain, ... , bolaphiles, gémini, ... (**Figure 1**).
- the length of the lipophilic part which makes it possible to classify surfactants as wetting agents (C8–C10), detergents (C12–C16), emulsifiers or softeners (C18–C22).
- their origin, natural or synthetic.

2. Anionic surfactants

2.1 Preparation and applications

These molecules have one or more functional groups that ionize in the aqueous phase to give negatively charged surfactant ions to which are associated cations such as alkali metals (Na, K) or quaternary ammoniums (NR_4). They have detergent, wetting and foaming properties and are therefore present in most hygiene and cleaning products. More rarely, they are used as emulsifiers. Among the anionic surfactants, we can count:

2.1.1 Carboxylate derivatives

Carboxylate-based surfactants are mainly used as soaps [7]. These are the salts of carboxylic acids, resulting from the saponification of triglycerides of vegetable (palm oil, olive oil, etc.) or animal (tallow) origin. The part hydrophilic consists of a saponified carboxylic group. These are products detergents, wetting agents and foamers. The saponification reaction leads to very alkaline solutions, a pH 10 can then be reached, resulting in irritant reactions which can appear on certain fragile skins following soaping.

Among the Carboxylate-based, we can count:

- **Sodium stearate** or sodium octadecanoate (**Figure 2**) prepared by saponification of triglycerides from vegetal or animal source. This sodium salt of stearic acid has many applications in scientific world as well as in daily use. This white solid is the most common soap. It is found in many types of solid deodorants, rubbers, latex paints, and inks.

- **Sodium laurate** or sodium dodecanoate (**Figure 3**) prepared by saponification of coconut oil or palm oil. In addition to its use in bar soaps and liquids, sodium laurate as a strong surfactant can be used at high concentrations to extract and solubilize the membrane proteins particularly integral membrane proteins.

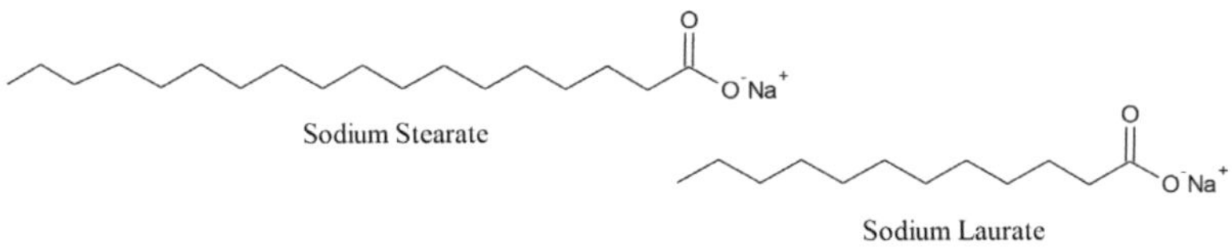

Figure 2.
Examples of carboxylate derivatives surfactants.

2.1.2 Amino acid carboxylate derivatives

Amino acid surfactants (AAS) are biocompatible and biodegradable surfactants obtained by condensation of natural amino acids with fatty acids (or their derivatives) of oleochemical source [8–10]. Moreover, AAS can be produced in large scale by green chemistry approaches, including enzyme-catalyzed synthesis using immobilized lipases and proteases, although chemical processes still prevail due to high yields and low production costs [11–16].

Among the Amino acid carboxylate -based, we can count:

- **Sodium lauroyl glycinate** (**Figure 3**) was used as the main surfactant in a body wash product for its mildness and lather properties [17].

- **Sodium lauroyl sarcosinate** (**Figure 4**), was widely used in personal care products and has immense biological and industrial significance. It is used in a number of cosmetic formulations and in soaps [18, 19].

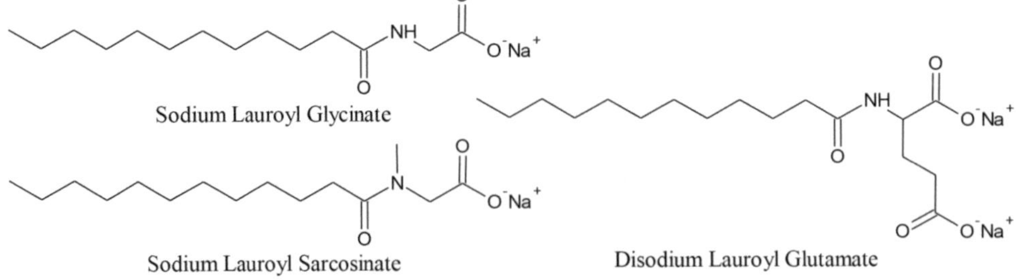

Figure 3.
Examples of amino acid carboxylate derivatives surfactants.

- **Disodium lauroyl glutamate** (**Figure 3**) have also been used in transparent soap bars [20] and this has been attributed to the presence of a unique cubic phase in between the micellar phase and the anisotropic hexagonal liquid crystalline phase in their phase behavior [21].

2.1.3 Sulphate derivatives

They are composed of a hydrophilic part consisting of a sulfate group (—O—SO_3^-) and a lipophilic alkyl chain, most often C12 and usually saturated (C12:0). They have the advantages of being good detergents, good foaming and wetting properties. Among them, we can count:

- Alkyl sulphates: sodium lauryl sulphate (SLS) or SDS (dodecyl sulphate of sodium) (**Figure 4**) are prepared by neutralization of the alkyl-ester-sulfuric acid by the appropriate base.

- Alkyl ether sulfates, (ex: Sodium Lauryl Ether Sulfate) (**Figure 4**) which are less irritating than SDS while having the same properties. For these surfactants, the ether group is a poly (ethylene oxide) chain (POE).

Same results obtained with sulfated alkalonamides prepared by sulfating of the alkanol amides, in particular those in C12–C14 (cocoamide). Neutralization of alkylamide sulfuric acid with monoethanolamine in the following example (**Figure 4**), resulting in a foam booster used in shampoos and bubble baths. These surfactants have a large hydrophilic group and do not irritate the skin. They are used as LSDA and foam stabilizers in soap bars and shampoos.

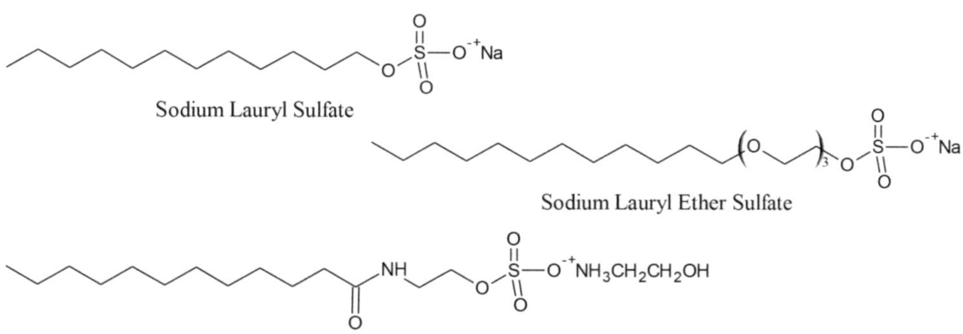

Figure 4.
Examples of Sulphate derivatives surfactants.

Petroleum sulfonates are sulfonates produced when an intermediate-molecular-weight refinery stream is sulfonated (**Figure 5**). These acids were then transformed into the corresponding surfactant and removed from the oil by liquid-liquid extraction with an alkaline solution (**Figure 5**). These surfactants have been widely used because they are effective at attaining low Interfacial tension, are relatively inexpensive, and have been reported to be chemically stable [22].

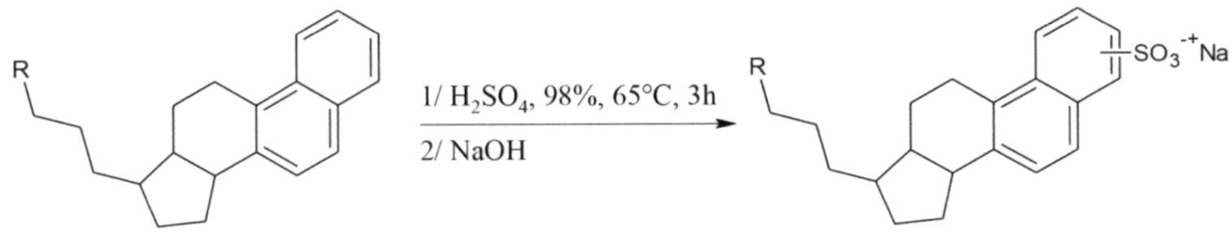

Figure 5.
General chemical reaction taking place in the preparation of petroleum sulfonates.

2.2 Phosphate ester surfactants

Alkyl phosphates are the organic esters of ortho-phosphoric acid. These ingredients are mixtures of esters and salts wherein a phosphate may have one to three alkylations and one to two potassium or sodium cations (**Figure 6**). Phosphate esters are anionic surfactants which are produced by phosphation of aliphatic or aromatic (ethoxylated) alcohols [23]. The properties of phosphate esters can be altered by the type of alcohol used as raw material and degree of ethoxylation of the alcohol [24]. Phosphate esters are known for exhibiting mild and nonirritating properties [25]. Phosphate esters are widely used as lubricants and hydraulic fluids because of high performance and safety advantage over others due to their high thermal stability [26].

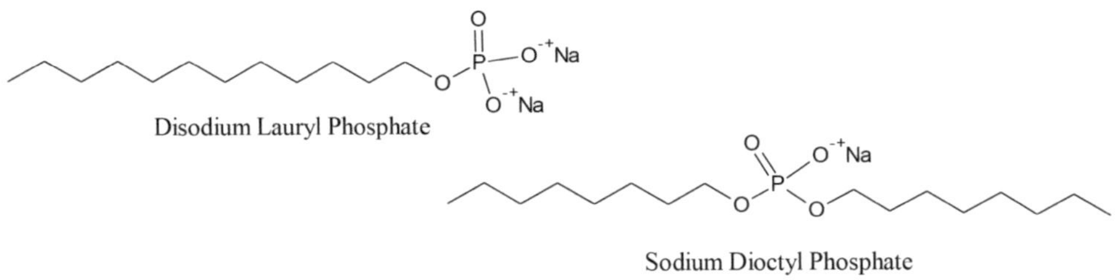

Figure 6.
Examples of alkyl and dialkyl phosphate esters.

3. Cationic surfactants

They contain a positively charged ionic group in aqueous solution. Their part hydrophilic positively charged allows them to adsorb very easily on substrates negatively charged. Thus, this positive charge gives them in particular affinities with the keratin, protein which mainly constitutes the hair (95%). On the other hand, many of them possess bactericidal properties. Indeed, they have the ability to destroy (lyse) the membrane of microorganisms when the fatty chain is short. From Cationic surfactants, the most used are quaternary ammonium and pyridinium derivatives.

3.1 Alkyl-ammonium surfactants

Alkyl-ammonium ions are produced in acid medium by the reaction of a proton with the amine (**Figure 7**). The resulting salt (in general chloride or bromide) is soluble in water thanks to the cation solvation. Fatty amines come from fatty acids, hence their chain is linear with a even number of carbon atoms. These surfactants are used as antistatic agent, emulsifying agent, and dispersing agent and corrosion inhibitor [27–29].

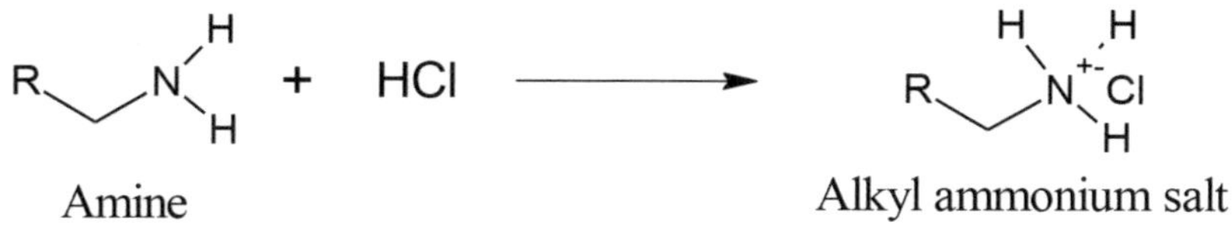

Figure 7.
General chemical reaction of preparation of ammonium salts.

3.1.1 Surfactants with one quaternary ammonium head

3.1.1.1 Quaternary ammonium salts

Quaternary ammonium salts also called Quats: are organic ammonium compounds to which the fourth valences of a nitrogen molecule are organically bound (**Figure 8**). In the general formula of quaternary ammonium salts, at least one of the radicals is formed by a long aliphatic chain of more than eight carbon atoms [30].

These products are stable in acidic and basic medium and are usually soluble in water and alcohols. They adsorb strongly on different materials, which can decrease the concentration of their solutions. Quaternary ammonium derivatives are bactericidal on a wide variety of germs, especially on Gram positive ones. They often have fungicidal, virucidal and algicidal properties. The pH, the presence of proteins, the hardness of the water influence their activities. Finally, the presence of carbon and nitrogen atoms in their molecules promotes the multiplication of resistant strains. Contamination of dilute solutions must therefore be avoided [31].

Quaternary ammonium salts are soluble both in aqueous medium because it is polar but also in organic medium since it has aliphatic chains: (R_4N^+, ^-X). They are used as phase transfer catalysis in a two-phase organic/aqueous medium [32–35].

3.1.1.2 Alkyls quaternary ammonium salts

Alkyl trimethylammonium is produced by the reaction of a fat-rich amine and chloromethane in the presence of sodium hydroxide. In the case of dialkyl dimethyl ammonium, the amine used is dialkylated. Alkyl-dimethyl benzyl quaternary ammonium is produced by the reaction of alkyl dimethylamine with benzyl chloride or epichlorohydrin. The number of carbons in the long alkyl chains in these surfactants is usually 8 to 22, while that of the short chains is 1 to 7 (**Figure 9**).

The most common quaternary ammonium disinfectants are probably benzalkonium chloride (BAC), and didodecyldimethylammonium chloride/bromide [36]. BAC is utilized in hand sanitizers, soaps, cleaning wipes, hospital sanitation kits, surface disinfectants [37, 38], etc. Quaternary ammonium compounds, with double-chain called dialkyl (e.g., didodecyldimethylammonium bromide (C12-C12DMA)), are used for emulsion stabilization [39].

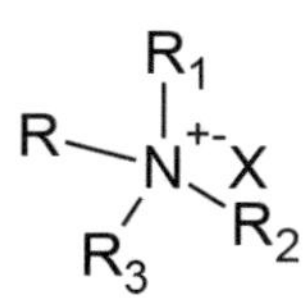

Figure 8.
Structure of an ammonium salt.

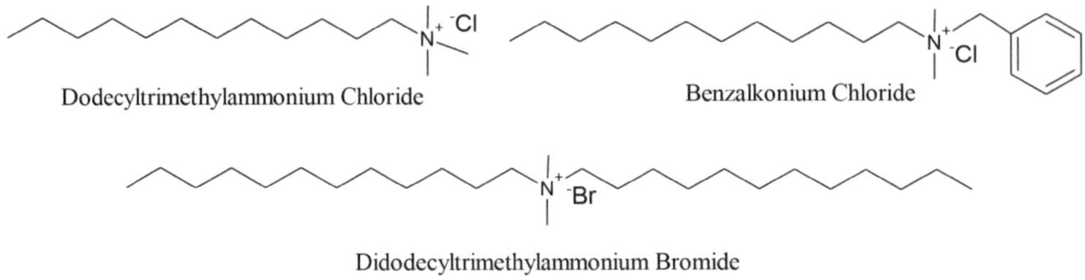

Figure 9.
Examples of alkyl trimethylammonium surfactants.

3.1.1.3 Alkylcyclic quaternary ammonium salts

Another important cationic surfactants class contains aromatic or saturated heterocycles including one or more nitrogen atoms. This is the case of a well used n-alkylpyridinium chloride, which is prepared by reacting alkylchlorure on pyridine. Pyridinium compounds (**Figure 10**) have been also used to achieve asymmetric and regioselective synthesis by additions of Grignard reagent [40]. Pyridinium cationic surfactants (**Figure 10**) have applications in the synthesis of TiO_2 nanoparticles [41], in the synthesis of ionic liquids [42] or as electrolytes for dye-sensitized solar cells [42]. Some of the pyridinium salts (C12 and C16) have been used as agents for solubilizing compounds in water in analytical chemistry applications [41].

Piperidinium compounds find application in hair conditioning formulations as antistatic agents. Piperidinium surfactants, including those functionalized with OH groups, show a high antimicrobial effect at concentrations lower than critical micelle concentration for both bacteria and fungi, with the low hemolytic activity revealed [43].

Piperidinium salts attract considerable interest due to their superior performance in terms of high ionic conductivity, wide electrochemical window, low volatility and low flammability. One of the most important applications of piperidinium salts is its application in electrolyte systems. They are used in battery systems [44, 45].

A cyclic compound with two nitrogen atoms, so-called imidazole is prepared from ethylene diamine [46]. Due to their low toxicity [47], these compounds are widely used as precursors in the synthesis of cationic surfactants called imidazolinium salts (**Figure 10**). Cationic imidazolines or imidazolium salts derive from N-alkylation which results in the attachment of an alkyl or aryl to the ring of the imidazoline.

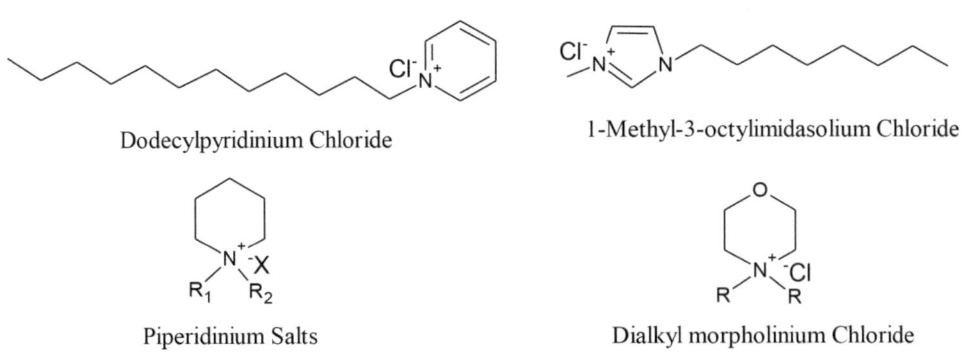

Figure 10.
Examples of alkylcyclic quaternary ammonium surfactants.

Imidazolium salts are much more hydrophilic than their bases and therefore function as stable acid detergents with good wetting agents [48]. They also show good lubricating power to form complexes with bentonites. They are also compatible with aqueous systems. The oleic acid salt of imidazolinium is a good lubricant, it is widely used in various industrial applications [48].

Morpholine is a saturated cycle containing both a oxygen and a nitrogen atom. The dialkylation of the nitrogen atom results in a salt called *N*,*N*-dialkyl-morpholinium. These dialkylation were observed when morpholine carried out with alkylhalogenur (**Figure 10**). Morpholinium cationic liquids are less toxic than other cationic groups of ionic liquids such as imidazolium, pyridinium, piperidinium, pyrrolidinium, etc. [49]. They were shown to perform well in the electrolyte systems for lithium batteries, supercapacitors and solar cells [50]. Morpholine and its N-Alkyl derivatives are a class of important intermediates for many fine chemicals, which have been widely used in rubber, medicine, pesticide and other fields (**Figure 10**) [51].

3.1.2 Surfactants with two quaternary ammonium heads

3.1.2.1 Gemini quaternary ammonium surfactants

Gemini surfactants, or dimers, are organic compounds formed from two hydrophilic heads and two hydrophobic chains, separated by a covalently bonded spacer [52–58].

Most studies, however reported on the surface tension of the aqueous solutions of gemini cationic surfactants for CMC determination and assessment of their capacity in the reducing the surface tension of water [59–61]. The CMC values of piperidinium and morpholinium gemini surfactants (**Figure 11**) are lower or comparable to counterparts reported in the literature such as conventional quaternary ammonium gemini surfactants and heterocyclic gemini surfactants. Many results of physicochemical property studies of gemini surfactants show that these properties depend not only on changing the length of the hydrophobic alkyl or the length of the spacer, but also on varying the type of cationic head group. They have found broad utility ranging from industrial to personal care applications [62–64]. When these salts with two alkyl chains and two ammonio groups show good extractive abilities as phase transfer

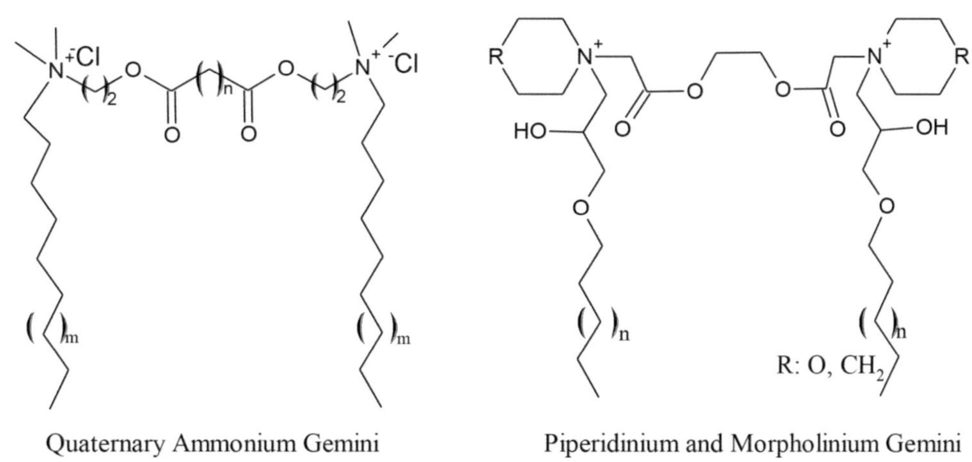

Figure 11.
Examples of gemini cationic surfactants.

catalyst [65]. These Cationic geminis surfactants were used to separate all members of a family ergot alkaloids by micellar electro kinetic capillary chromatography when the spacer is 1, 3-bis (dodecyl-*N,N*-dimethyl ammonium)-2-propanol dichloride [66].

3.1.2.2 Bolaphils quaternary ammonium surfactants

Bolaamphiphiles are composed of two polar head groups separated by one, two or three long hydrophobic spacers that are mostly alkyl chains, and can even be steroids or porphyrins [67]. Bolaamphiphiles can be divided in two main categories (**Figure 12**): symmetric [68] (with same polar headgroups at both ends) and asymmetric [69] ones (possessing different polar headgroups at both ends), however both capable of self assembling to yield interesting nano-assemblies. Chemical structure of bolaamphiphiles can dictate their arrangement into parallel or antiparallel sheets resulting in formation of either unsymmetrical or symmetrical monolayer membranes [70–72]. Bolaamphiphiles have been used in formulating stable nanocarriers systems and have already demonstrated enough potential in effective drug as well as gene delivery [73, 74].The self-assembly characteristics in aqueous solutions of cationic bolaamphiphiles with systematic changes in their chemical structure has been investigated [75]. Generally, it has been shown that symmetrical bolaphiles can self associate into mechanically very stable monolayer. Upon sonication, the single-chain bolaphile forms spherical aggregates [75].

Others formed a gelatinous network of fibers or multi-lamellar vesicles [76]. Vesicles made from cationic bolaamphiphiles (bolas), GLH-19 and GLH-20, (**Figure 13**) are used to investigate their ability to protect siRNAs from nuclease degradation while delivering them to target cells, including including brain cells are computational and experimental studies. These studies show that that GLH-19 vesicles have better delivery characteristics than do GLH-20 vesicles in terms of stability, binding affinity, protection against nucleases, and transfection efficiency, while GLH-20 vesicles contribute to efficient release of the delivered siRNAs, which become available for silencing [77–79]. Compared with vesicles prepared from phospholipids or double chain amphiphiles, the catanionic vesicles can form spontaneously and remain stable for a longer time, which takes the big advantage in practical applications [80].

The vesicles formed from cationic bolaamphiphile synthesis from *N,N′*-alkylenebis (vernolamide) and the corresponding α,ω-alkylene vernoldiester with choline esters as cleavable head groups (**Figure 13**) can be use for targeted drug delivery as hydrolysis of the head group by specific enzyme, abundant in the target organ, will results in efficient decapsulation of the vesicles and the release of the encapsulated drug at the target site [81]. These bolaphiles have much weaker surface activity but stronger aggregation ability which is characterized by much higher surface tension but lower CMC (critical micelle concentration) than their conventional counterparts with the same hydrocarbon/head group ratio [82–84].

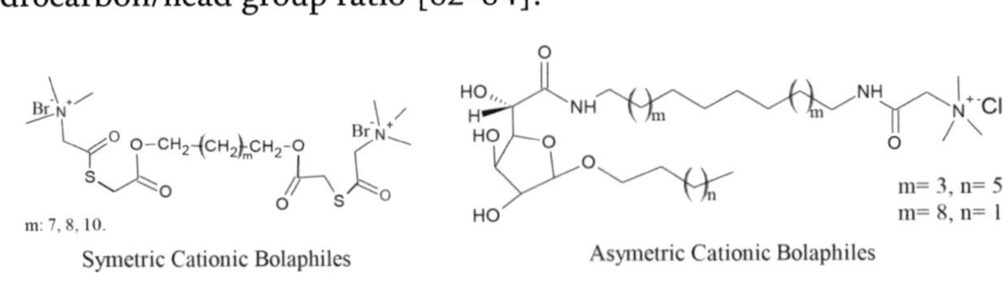

Figure 12.
Examples of symmetric and asymmetric cationic surfactants.

4. Nonionic surfactants

Nonionic surfactants contain an uncharged polar group with a high affinity to water. They consist of a saturated or slightly unsaturated hydrocarbon chain to which are attached several polar groups such as ether, alcohol, carbonyl, amine type or a polymeric chain, in particular a polyoxyethylene chain.

4.1 Ethoxylated nonionic surfactant

4.1.1 Alcohol ethoxylates

Alcohol ethoxylates (EAs) are a class of nonionic surfactants that contain a hydrophobic alkyl chain attached to a chain of hydrophilic ethylene oxide (EO) groups through an ether bond, and have a general structure $R(OCH_2CH_2)$ nOH (**Figure 14**). These products are based on either synthetic or natural fatty alcohols [85]. Synthetic alcohol ethoxylates are produced by direct ethoxylation of alcohols, whereas natural fatty alcohols are first reduced to make them saturated before going for ethoxylation. Both synthetic and natural alcohol ethoxylates are used as nonionic surfactants in many industries. They are mainly used in industrial and consumer products, such as laundry detergents and multi-purpose cleaning products; to a lesser extent, they are also used by the agriculture, cosmetics, textiles, paper and petroleum products sectors [86, 87].

Lauryl alcohol ethoxylate, stearyl alcohol ethoxylate, behenyl alcohol ethoxylate, etc. are examples of fatty alcohol ethoxylates. All these ethoxylated products vary in physical appearance and have different properties like pour point, cloud point, density, viscosity, and flash point depending on the level of ethoxylation process from which they are formed [23].

4.1.2 Fatty acid ethoxylates

Fatty acid ethoxylates are produced by ethoxylation of fatty acids derived from saturated or unsaturated animal or vegetable fats like Coconut Fatty Acid, Lauric

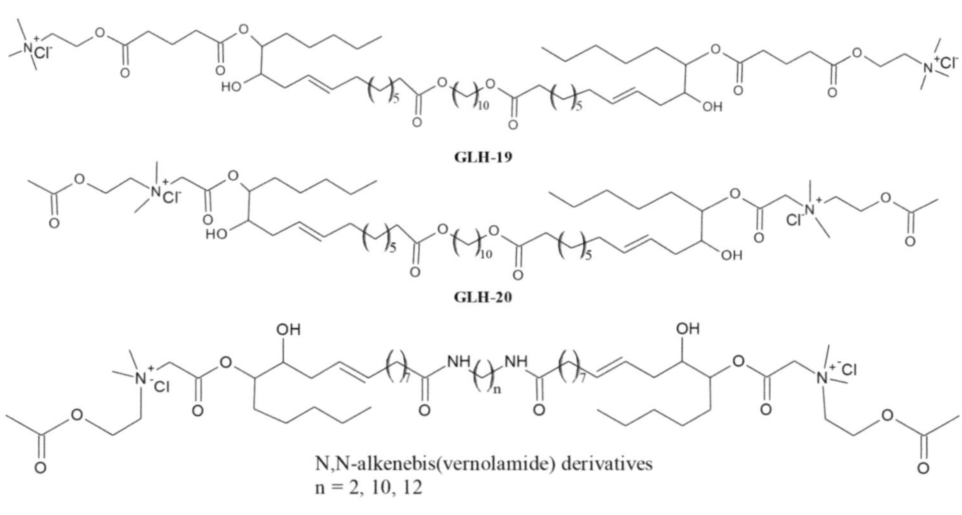

Figure 13.
Chemical structure of selected bolaamphiles GLH-19, 20, and N,N′-alkylenebis(vernolamide) derivatives.

Acid, Oleic Acid and Stearic Acid. These nonionic surfactants manufactured by addition of ethylene oxide to fatty acids producing a range of ethoxylates having different moles of EO. These are extensively used in formulations of emulsifying softeners, wetting agents, cleansing agents and dispersants. The stearic acid ethoxylates (**Figure 15**) are applied in cosmetic industrial sectors as emulsifiers in oil-in-water type creams and lotions. Fatty acid ethoxylate includes; lauric acid ethoxylates and coco fatty acid ethoxylate. These products are used in textile and in the manufacture of spin finishes [23].

4.1.3 Alkyl phenol ethoxylates

The basic element of the alkylphenol molecule is a phenolic ring on which is substituted, generally in the para position, an octyl, nonyl, or dodecyl radical (**Figure 16**). This molecule comprises a variable number of ethoxylate groups ($—CH_2—CH_2—O—$) in the —OH functional group of phenol. The number of ethoxylates determines the physico-chemical properties of the product and is fixed according to the uses for which it is intended. Most common commercially used products are octyl, nonyl, and dodecylphenol with a degree of ethoxylation ranging from 4 to 40. Octyl and nonylphenols with 8–12 ethylene oxide moles are used in detergents [23]. Ethoxylates with ethylene oxide moles ranging from 12 to 20, are wetting agents and O/W emulsifiers. More than 20 ethylene oxide moles exhibit detergents properties at high temperature and high salinity [23]. Compounds having ethylene oxide moles <5 are applied in antifoaming agents or detergent in non-aqueous media [23].

4.1.4 Cyclic polyoxyethylene alkyl ethers

The attachment of lipophilic long-chain alkyl groups allows crown ethers to acquire some amphiphilic properties [88, 89]. When dispersed in an aqueous medium, these molecules self-organize upon sonication, often into monolayer vesicles [90].

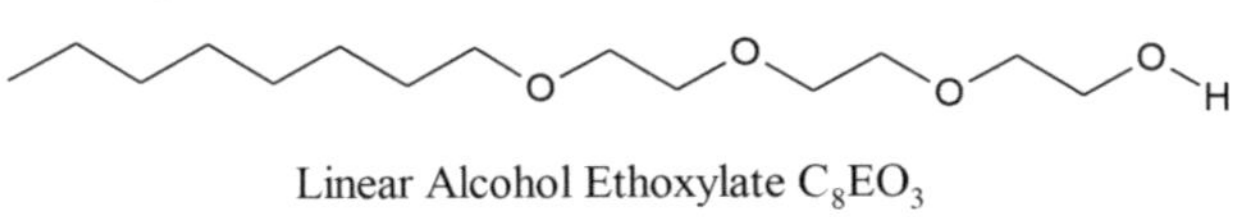

Figure 14.
Chemical structure of alcohol ethoxylates.

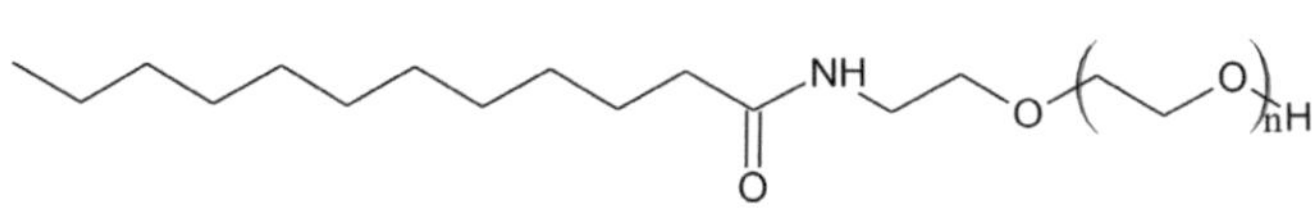

Figure 15.
Chemical structure of fatty acid ethoxylates.

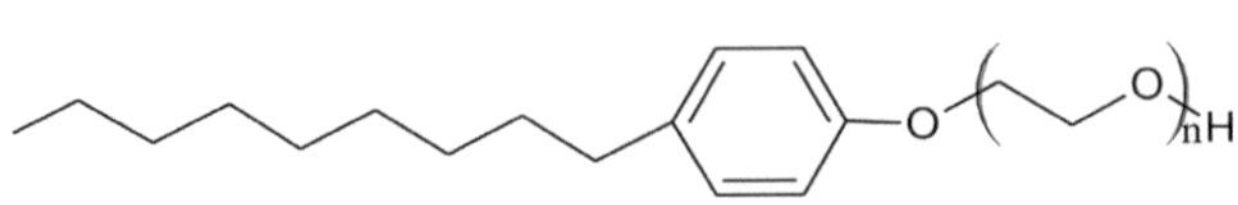

Figure 16.
Chemical structure of nonylphenol ethoxylates.

Nevertheless, the use of fluorocarbon chains instead of hydrocarbon analogues seems to constitute an efficient way to stabilize vesicular systems [91, 92] and to achieve their potential as drug delivery systems [93, 94].

Cyclic polyoxyethylene dialkyl ethers [95] are obtained by grafting two alkyl chains onto the crown head which is capable of organizing into membranes [90, 96]. These surfactants (**Figure 17**) may be inefficient in the formation of aggregates because of their low hydrophilicity. This problem is solved by the addition of a cap anion [97] or by the introduction of an oligooxyethylene group between the hydrophobic part and the ring of crown ethers [98–100] to reinforce the hydrophilicity.

4.1.5 Bolaphiles cyclic polyoxyethylene surfactants

Crown ether bolaphiles are molecules comprising two polar heads joined by one or more hydrophobic chains. Crown ethers based srfactant have been studied in various stages of drug delivery. The results of their studies showed that bolaform compounds BD-16 and BC-16 (**Figure 18**) are able to form niosomes, if prepared according to the thin-film hydration method, in which hydrophilic molecules can be entrapped [101]. They evaluated its ability to form vesicles and the influence of a membrane additive, such as cholesterol. Bola A-16 was used as the niosome carrier for 5-fluorouracil (5-FU).

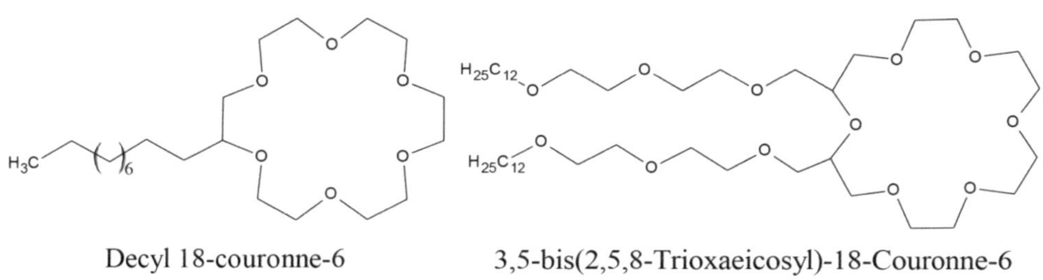

Figure 17.
Examples of Crown ethers surfactants.

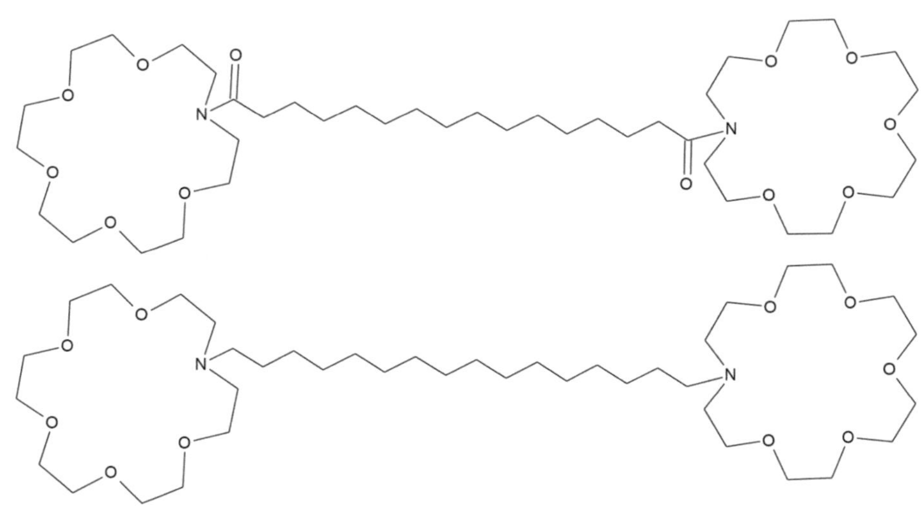

Figure 18.
Chemical structure of BD-16 and BC-16.

The results were very interesting and showed that the stability of the drug in encapsulated state in niosomes is more than a free solution [102].

4.2 Polyhydroxyl (polyol)-based surfactants

4.2.1 Sugar-based surfactants

Single-chain surfactants synthesized from sugars occur in large majority in the form of alkylpolyglycosides (APG) (**Figure 19**). They result from the fixation by O-glycosylation of an alkyl chain of variable length (8 to 18 carbons) on a unit carbohydrate in its pyranose or furanose form, usually glucose [103, 104], galactose [105] or more disaccharides.

The hydrophilic part of alkyl polyglucosides (APG) derives from sugars and the hydrophobic part comes from a natural or synthetic source, or from a mixture of fatty alcohols. Both monomeric and polymeric sugars can be used as feedstock [106]. The alkyl chain of APGs depends on the raw material used: coconut oil and palm kernel oil are used for the C12/C14 range, while tallow and rapeseed oil for the C16/C18 chains [107].

During the direct synthesis of APGs, glucose is suspended in an excess of fatty alcohol and the reaction takes place at 100–120°C, in the presence of an acid catalyst such as sulfonic acid. After removal of water, which is a by-product of the reaction, the final product is composed of a mixture of, alkyl oligoglucosides and alkyl polyglucosides. The average degree of polymerization for this type of APGs depends on the glucose/alcohol ratio [104].

The biodegradability study showed that APGs with the chain lengths of C8–C10, C10–C12, C12–C14 decompose from 95% to 100%. The toxicity study confirmed their compatibility with the environment and their less irritating nature than conventional surfactants [108].

4.2.2 Glycosides of fatty acids

The Simple glycolipids (**Figure 20**) are composed of a glycosyl group (one or more units) linked to a hydroxyl fatty acid or to a carboxyl group of a fatty acid (ester bond). These compound are found in bacteria, yeasts and lower marine invertebrates. They frequently possess interesting physical or biological properties [104, 107–111].

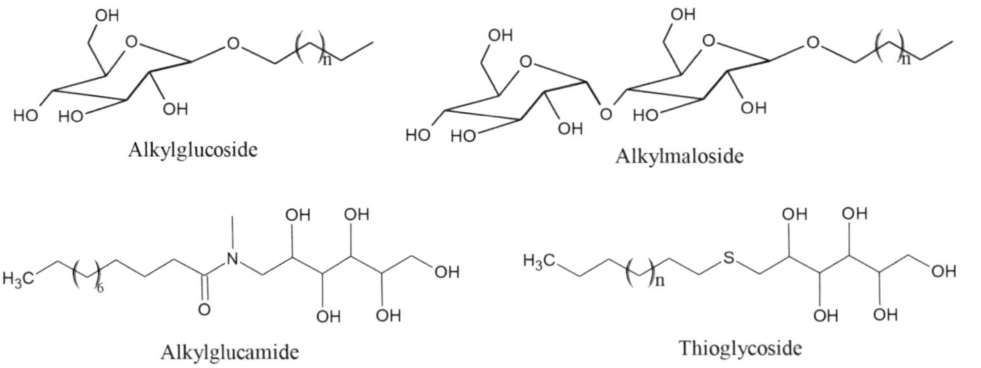

Figure 19.
Chemical structure of monotailed sugar-based surfactants.

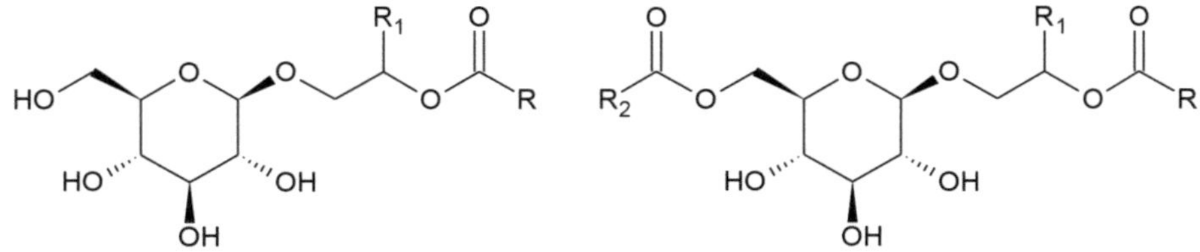

$R = C_{13}H_{27}$, $R_1 = C_{17}H_{35}$.
$R = C_{15}H_{31}$, $R_1 = C_{18}H_{37}$.

$R = R_2 = C_{13}H_{27}$, $R_1 = C_{18}H_{37}$.
$R = R_2 = C_{15}H_{31}$, $R_1 = C_{18}H_{37}$.

Figure 20.
Examples of bi-tailed and tri-tailed glycosides based surfatants.

4.2.3 Sugar-based bolaamphiphiles

Sugar-based bolaamphiphiles [96, 112–114] are of special interest for pharmaceutical and other applications because of their natural origin, biocompatibility and biodegradability. A large variety of non-ionic [111, 112] sugar-based bolaamphiphiles (**Figure 21**) have been reported so far with wide variety of properties and applications, e.g., formulation of anti-HIV com bination therapy [111], anticancer drug delivery [112], gene transfection [113], bacterial agglutination [114], etc.

4.2.4 Sorbitan esters of fatty acids

Sorbitan esters are nonionic surfactants that are commonly used as water-in-oil (w/o) emulsifies and stabilizes. Sorbitan fatty acid esters are produced by esterification of dehydrated sorbitol (sorbian) with a varity of fatty acids. There are many types of sorbitan esters with different kinds of fatty acids and varying degrees of esterification (**Figure 22**). Sorbitan esters are often used in the food industry, especially in chocolate and margarine production, where a high affinity for fats is desired. Polysorbates are also commonly used as emulsifiers in the food industry (salad dressing, ice cream) and in cosmetics, pharmaceuticals, detergents, paints, and plastic applications.

Polysorbate, a substance prepared by the reaction of sorbitan fatty acid ester (a nonionic surfactant) with ethylene oxide. Polysorbates ("tweens") are widely used in industry and medicine as antistatic agents, defoamers, emulsifiers and solubilizers, fiber softeners, dispersing agent and substance that contributes to the dissolution of both water-soluble in the fat phase and waxes in the pharmaceuticals, and fat-soluble in the aqueous phase, in animal feed and soil treatment [115–118].

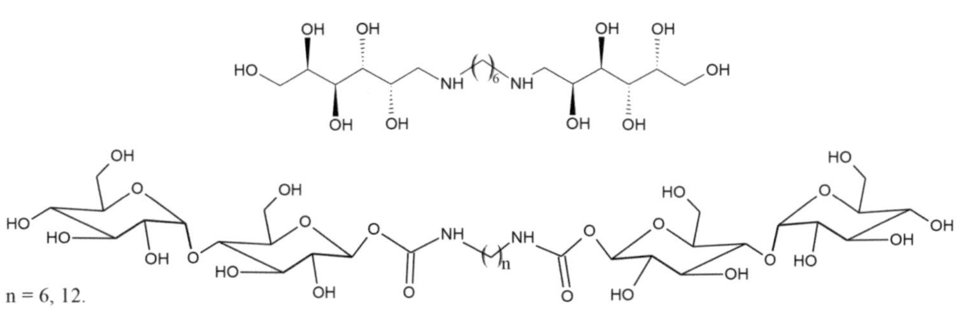

Figure 21.
Examples of sugar-based bolaphiles.

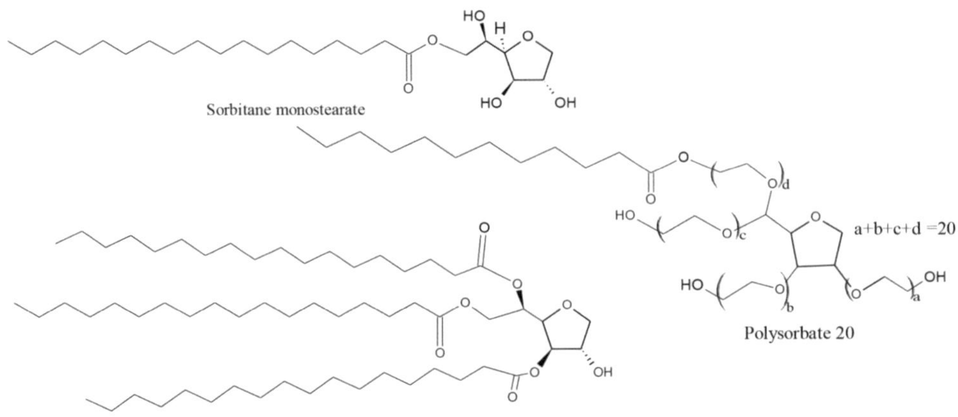

Figure 22.
Examples of Sorbitan based surfactants.

5. Amphoteric or zwitterionic surfactants

Amphoteric surfactants have two functional groups, one anionic and the other cationic. In most cases, it is the pH that determines the dominant character. This dipolar structure is similar to that of natural phospholipids and leads to a family products that are generally non-irritating and not very aggressive on the biological level. They are compatible with both anionic surfactants and cationic surfactants and present the main properties of these two categories. Thus, they are detergents, wetting, foaming and slightly bactericidal. The main amphoteric surfactants are betaine derivatives (alkybetaine) and imidazoline derivatives (amphoacetates).

5.1 Phospholipids

Phospholipids are esters of glycerol fatty acids with a phosphoester group linking the polar part to the hydrophobic composed of two aliphatic chains (**Figure 23**). According to the nature of the two pole parts and apolar, there is a wide variety of phospholipids, the most common of which are phosphatidylcholines, sphingomyelins and phosphatidylserines. Structurally, all phospholipids are amphiphilic molecules, serving as the molecular basis for their biological functions and as emulsifiers in food, cosmetic, or drug products.

Phospholipids are major membrane lipids that consist of lipid bilayers (**Figure 24**). This basic cellular structure acts as a barrier to protect the cell against various environmental insults and more importantly, enables multiple cellular processes to occur

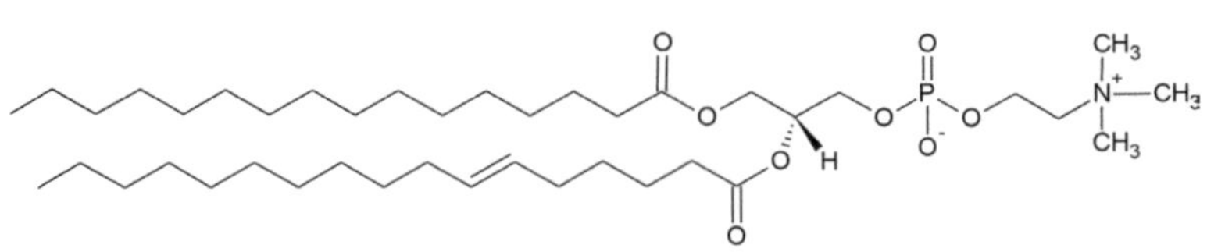

Figure 23.
Phosphatidylcholine.

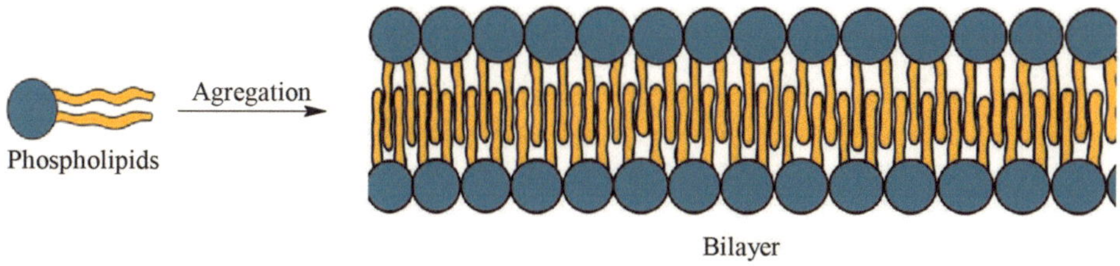

Figure 24.
Agegation of phopholipids to bilayer.

in subcellular compartments. PLs also function as precursors of second messengers such as diacylglycerol (DG) and inositol-1,4,5-P_3 (inositol trisphosphate). A third, and usually overlooked, function of PLs is the storage of energy in the form of fatty acyl components [119].

5.2 Alkyl betaines

Dodecylbetaine chloride (C12Be) has been synthesized and purified by different methods. One method is to react sodium chloroacetate with *N*,*N*-dimethylamine-*N*-dodecylamine (molar ratio 1:1) in an aqueous solution of ethanol under reflux for 18 h [120, 121]. Alkyl betaine (**Figure 25**) is a very mild detergent that leads to a large amount of foam with good sensory properties. It is therefore logically used in a large number of hygiene formulations.

Lauryl betaine was reported to be used in hair sprays, body and hand products, non-coloring hair powders, and indoor tanning preparations that may be aerosolized or become airborne and could possibly be inhaled.

Zwitterionic surfactants based on 3-(1-alkyl-3-imidazolio) propane-sulfonate [ImS3-R] is an emerging and important class of amphiphile due to their relevance as nano reactors for the synthesis of metallic nanoparticles and accelerated acid hydrolysis [122].The solubility of these surfactants is very low but increases when adding a salt which helps to stabilize the micellized surfactant, an increase in temperature increased the surface activity and decreased the CMC of both zwitterionic surfactants, although the changes were small [122].

Electrophoretic evidence and fluorescence inhibition for ImS3-14 (**Figure 26**) shows that the micellar aggregation number is only slightly sensitive to added salts, as is the critical micelle concentration, but $NaClO_4$ markedly increases zeta potentials of ImS3-14 in a similar way as in *N*-tetradecyl-*N*,*N*-dimethylammonio-1-propanesulfonate (SB3-14) micelles (**Figure 26**) [123].

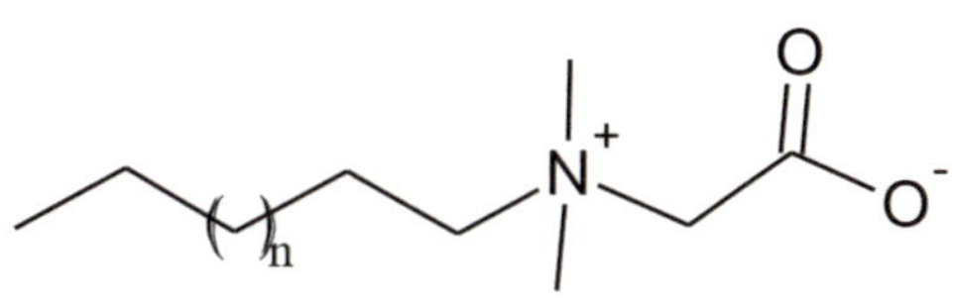

Figure 25.
Alkyl betaines.

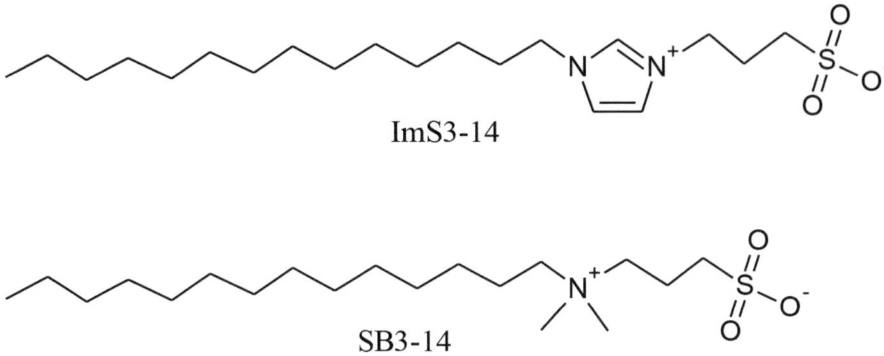

Figure 26.
Structure of ImS3-14 and SB3-14.

6. Conclusion

In this study, we have shown that surfactants play an important role in various areas of daily life. Also that these have applications, depending on their structures and the conditions in which they are found. It is therefore possible to imagine that an ideal surfactant remains, in reality, inaccessible in any area whatsoever.

Author details

Ali Khalfallah[1,2]

1 Faculty of Sciences of Gabes, Gabes, Tunisia

2 Laboratory of Molecular Organic Chemistry, National Higher Engineering School of Tunis, Tunis, Tunisia

*Address all correspondence to: ali.khalfallah@fst.rnu.tn

References

[1] Rosen MJ. Surfactants and Interfacial Phenomena. 3rd ed. New York: John Wiley & Sons; 2004. p. 464

[2] Ho Tan Tai L. Détergents et produits de soins corporels, Dunod. Paris; 1999

[3] Bagha-Bagha AR, Holmberg K. Solubilization of hydrophobic dyes in surfactant solutions. Materials. 2013;**6**: 580-608

[4] Schramm L, Stasiuk E, Marangoni D. Annual reports on the progress of chemistry. Section C: Physical Chemistry. 2003;**99**:3-48

[5] Griffin WC. Classification of surface-active agents by "HLB". Journal of the Society of Cosmetic Chemists. 1949;**1**: 311-324

[6] Davies JT, Rideal EK. Interfacial Phenomena. 2nd ed. New York: Academic Press; 1963. pp. 371-374

[7] Crepaldi E, Pavan P, Tronto J, Valim J. Carboxylate surfactants as efficient and renewable promoters for methane hydrate formation. Journal of Colloid Interface Science. 2002;**248**(2):429-442

[8] Pinazo A, Pons R, Pérez L, Infante MR. Amino acids as raw material for biocompatible surfactants. Industrial and Engineering Chemistry Research. 2011; **50**:4805-4817

[9] Infante MR, Pérez L, Pinazo A, Clapés P, Morán MC, Angelet M, et al. Amino acid-based surfactants. Comptes Rendus Chimie. 2004;**7**:583-592

[10] Faustino CMC, Serafim CS, Ferreira IN, Branco MA, Calado ART, Garcia-Rio L. Mixed micelle formation between an amino acid-based anionic gemini surfactant and bile salts. Industrial and Engineering Chemistry Research. 2014; **53**:10112-10118

[11] Morán MC, Pinazo A, Pérez L, Pinazo A, Clapés P, Angelet M, et al. "Green" amino acid-based surfactants. Green Chemistry. 2004;**6**:233-240

[12] Villeneuve P. Lipases in lipophilization reactions. Biotechnology Advances. 2007;**25**:515-536

[13] Morcelle SR, Liggieri CS, Bruno MA, Priolo N, Clapés P. Screening of plant peptidases for the synthesis of arginine-based surfactants. Journal of Molecular Catalysis B: Enzymatic. 2009;**57**:177-182

[14] Reznik GO, Vishwanath P, Pynn MA, Sitnik JM, Todd JJ, Wu J, et al. Use of sustainable chemistry to produce an acyl amino acid surfactant. Applied Microbiology and Biotechnology. 2010; **86**:1387-1397

[15] Franssen MCR, Steunenberg P, Scott EL, Zuilhofac H, Sanders JPM. Immobilised enzymes in biorenewables production. Chemical Society Reviews. 2013;**42**:6491-6533

[16] Zhang G, Xu B, Han F, Zhou Y, Liu H, Li Y, et al. Green synthesis, composition analysis and surface active properties of sodium cocoyl glycinate. American Journal of Analytical Chemistry. 2013;**4**:445-450

[17] Regan J, Mollica L, Ananthapadmanabhan KP. A novel glycinate-based body wash. The Journal of Clinical and Aesthetic Dermatology. 2013;**6**:23-30

[18] Lanigan RS. Final report on the safety assessment of Cocoyl Sarcosine, Lauroyl Sarcosine, Myristoyl Sarcosine,

Oleoyl Sarcosine, Stearoyl Sarcosine, Sodium Cocoyl Sarcosinate, Sodium Lauroyl Sarcosinate, Sodium Myristoyl Sarcosinate, Ammonium Cocoyl Sarcosinate, and Ammonium Lauroyl Sarcosinate. International Journal of Toxicology. 2001;**20**(Suppl. 1):1-14

[19] Wallach DFH, Mathur R, Redziniak GJM, Tranchant JF. Some properties of N-acyl sarcosinate lipid vesicles. Journal of Social Cosmetic Chemists. 1992;**43**: 113-118

[20] Friedman M. Soap Manufacturing Technology. 2nd ed. AOCS; 2016. pp. 73-106

[21] Kaneko D, Olsson U, Sakamoto K. Self-assembly in some N-lauroyl-l-glutamate/water systems. Langmuir. 2002;**18**:4699-4703

[22] Green DW, Willhite GP. Enhanced oil recovery. In: Society of Petroleum Engineers Textbook. Vol. 6. Houston, TX, USA: Soc. Petroleum Eng.; 1998. p. 545

[23] Derian PJ, Tao G, Herve PJC, Reierson RL. High Solids, Pumpable Aqueous Compositions of High Monoalkyl Phosphate Ester Salt Content, U.S. Patent 6262130. 2001

[24] Arora P, Singh R, Seshadri G, Tyagi AK. Synthesis, properties and applications of anionic phosphate ester surfactants: A review. Tenside, Surfactants, Detergents. 2018;**55**(4): 266-272

[25] Williamson SC. Cleansing Compositions Comprising Ethoxylated Alcohol Monoesters of Phosphoric Acid, U.S. Patent 4493782. 1985

[26] Johnson DW, Hils JE. Phosphate esters, thiophosphate esters and metal thiophosphates as lubricant additives. Lubricants. 2013;**1**:132-148

[27] Lee J, Peacor D. Expansion of smectite by laurylamine hydrochloride: Ambiguities in transmission electron microscope observations. Clays and Clay Minerals. 1986;**34**(1):69-73

[28] Olajire A. Review of ASP EOR (alkaline surfactant polymer enhanced oil recovery) technology in the petroleum industry: Prospects and challenges. Energy. 2014;**77**:963-982

[29] Schramm L. Surfactants. Cambridge: Cambridge University Press; 2000. ISBN 0521640679

[30] Lever M, Atkinson W, George PM, Chambers ST. Abnormal urinary excretion of glycine betaine may persist for years. Clinical Biochemistry. 2007; **40**:798-801

[31] Likes R, Madl RL, Zeisel SH, Craig SAS. The betaine and choline content of a whole wheat flour compared to other mill streams. Journal of Cereal Science. 2007;**46**:93-95

[32] Caubère P. La catalyse par transfert de phase et son utilisation en chimie organique. Masson; 1982

[33] Lundberg D, Ljusberg-Wahren H, Norlin A, Holmberg KJ. Studies on dodecyl betainate in combination with its degradation products or with phosphatidyl choline–phase behavior and hemolytic activity. Journal of Colloid and Interface Science. 2004;**278**(2): 478-487

[34] Mohlin K, Karlsson P, Holmberg K. Use of cleavable surfactants for alkyl ketene dimer (AKD) dispersions. Colloids and Surfaces, A: Physicochemical and Engineering Aspects. 2006;**274**(1):200-210

[35] Tehrani-Bagha A-R, Bahrami H, Mouvassagh B, Arami M, Amirshahi SH, Menger FM. Dynamic adsorption of gemini and conventional cationic surfactants onto polyacrylonitrile. Colloids and Surfaces, A: Physicochemical and Engineering Aspects. 2007;**307**:121-127

[36] Ogilvie BH, Solis-Leal A, Lopez JB, Poole BD, Robinson RA, Berges BK. Alcohol-free hand sanitizer and other quaternary ammonium disinfectants quickly and effectively inactivate SARS-CoV-2. The Journal of Hospital Infection. 2021;**108**:142-145

[37] Vereshchagin AN, Frolov NA, Egorova KS, Seitkalieva MM, Ananikov VP. Quaternary Ammonium Compounds (QACs) and Ionic Liquids (ILs) as Biocides: From simple antiseptics to tunable antimicrobials. International Journal of Molecular Sciences. 2021;**22**:6793-6793

[38] Schrank CL, Minbiole KPC, Wuest WM. Are quaternary ammonium compounds, the workhorse disinfectants, effective against severe acute respiratory syndrome-coronavirus-2? ACS Infectious Diseases. 2020;**6**:1553-1557

[39] Eastoe J, Hetherington KJ, Dalton JS, Sharpe D, Lu JR, Heenan RK. Microemulsions with didodecyldimethylammonium bromide studied by neutron contrast variation. Journal of Colloid and Interface Science. 1997;**190**:449-455

[40] Lavilla R. Recent developments in the chemistry of dihydropyridines. Journal of the Chemical Society, Perkin Transactions. 2002;**1**(9):1141-1156

[41] Ardizzone S, Bianchi C, Cappelletti G. Growth of TiO2 nanocrystals in the presence of alkylpyridinium salts: the interplay between hydrophobic and hydrophilic interactions. Surface and Interface Analysis. 2006;**38**:452-457

[42] Deng F-G, Hu B, Sun W, Chen J, Xia CG. Novel pyridinium based cobalt carbonylionic liquids: Synthesis, full characterization, crystal structure and application in catalysis. Dalton Transactions. 2007;**38**:4262-4267

[43] Kushnazarova RA, Mirgorodskaya AB, Kuznetsov DM, Tyryshkina AA, Voloshina AD, Gumerova SK, et al. Modulation of aggregation behavior, antimicrobial properties and catalytic activity of piperidinium surfactants by modifying their head group with a polar fragment. Journal of Molecular Liquids. 2021;**336**(15):116318-116319

[44] Baranchugov V, Markevich E, Pollak E, Salitra G, Aurbach D. Amorphous silicon thin films as a high capacity anodes for Li-ion batteries in ionic liquid electrolytes. Electrochemistry Communications. 2007;**9**(4):796-800

[45] Salminena J, Papaiconomou N, Kumar RA, Leea J-M, Kerr J, Newman J, et al. Physicochemical properties and toxicities of hydrophobic piperidinium and pyrrolidinium ionic liquids. Fluid Phase Equilibria. 2007;**261**(1-2):421-426

[46] Tyagi R, Tyagi VK, Pandey SK. Imidazoline and Its Derivatives: An Overview. 2007;**56**(5):211-222

[47] Vorbruggen H, Krolikiewiz K. A simple synthesis of Δ^2-oxazolines, Δ^2-oxazines, Δ^2-thiazolines and Δ^2-imidazolines. Tetrahedron Letters. 1981;**22**:4471-4474

[48] Bajpai D, Tyagi VK. Fatty imidazolines: Chemistry, synthesis, properties and their industrial applications. Journal of Oleo Science. 2006;**55**(7):319-329

[49] Kumari M, Singh UK, Singh P, Pate R. Effect of N-butyl-N-methyl-Morpholinium bromide ionic liquid on the conformation stability of human serum albumin. Chemistry Select. 2017;**2** (3):1241-1249

[50] Pernak J, Borucka N, Walkiewicz F, Markiewicz B, Fochtman P, Stolte S, et al. Synthesis, toxicity, biodegradability and physicochemical properties of 4-benzyl-4-methylmorpholinium-based ionic liquids. Green Chemistry. 2011;**13**: 2901-2910

[51] Zhu B, Zhang Q, Shi S, Ye X, Qian C, Chen X. Overview on the synthesis of morpholine and its N-alkyl derivatives. Hans Journal of Chemical Engineering and Technology. 2017;**7**(1):31-37

[52] Menger F, Littau CA. Gemini-surfactants: Synthesis and properties. Journal of the American Chemical Society. 1991;**113**:1451-1452

[53] Wettig SD, Nowak P, Verrall RE. Thermodynamic and aggregation properties of gemini surfactants with hydroxyl substituted spacers in aqueous solution. Langmuir. 2002;**18**:5354-5359

[54] Wettig SD, Li X, Verrall RE. Thermodynamic and aggregation properties of gemini surfactants with ethoxylated spacers in aqueous solution. Langmuir. 2003;**19**:3666-3670

[55] Groth C, Nyden M, Holmberg K, Kanicky JR, Shah DO. Kinetics of the self-assembly of gemini surfactants. Journal of Surfactants and Detergents. 2004;**7**:247-256

[56] Fan H, Meng L. Superior thermal stability gel emulsion produced by low concentration gemini surfactant. Colloids and Surfaces A: Physicochemical and Engineering Aspects. 2011;**384**(1-3):194-199

[57] Fatma N, Panda M, Ansari WH, Kabir-ud-din. Solubility enhancement of anthracene and pyrene in the mixtures of a cleavable cationic gemini surfactant with conventional surfactants of different polarities. Colloids and Surfaces A: Physicochemical and Engineering Aspects. 2015;**467**:9-17

[58] Bhadani A, Shrestha RG. Self-aggregation properties of new ester-based gemini surfactants and their rheological behavior in the presence of cosurfactant — monolaurin. Colloids and Surfaces A: Physicochemical and Engineering Aspects. 2014;**461**:258-266

[59] Devinsky F, Masarova L, Lacko I. Surface activity and micelle formation of some new bis-quaternary ammonium salts. Journal of Colloid and Interface Science. 1985;**105**:235-238

[60] Devinsky F, Lacko I. Surface activity and micelle formation of some bis-quaternary ammonium salts of glycerine derivative. Tenside, Surfactants, Detergents. 1990;**27**:334-349

[61] DevinskY F, Lacko I, Bitterova FB, Tomeckova L. Relationship between structure, surface activity, and micelle formation of Some new bisquaternary isosteres of 1,5-pentanediammonium dibromides. Journal of Colloid and Interface Science. 1986;**114**:314-322

[62] Kwetkat K. Betaine Gemini Surfactants Made From Amines WO 9731890. p. 1997

[63] Steiger PH, Clansy SF, Portwood O. Diquaternary Compounds Useful as Bleach Activators, and Compositions Containing Them, U.S. Patent 5662827. 1997

[64] Keys RO, Friedli FE, Dalrymple DM, Manning M, Poffenberger C, Whittlinger DE, Hou W. Polyester

Polyquaternary Compounds, Compositions Containing Them, and Use Thereof, U.S. Patent, 6,211,139. 2001

[65] Lissel M, Feldman D, Nir M, Rabinovitz M. Bis-quaternary ammonium salts as phase transfer catalysts. Tetrahedron Lett. 1989;**30**: 1683-1686

[66] Menger FM, Keiper JS. Surface properties of novel surfactant, dihexadecyl gemini phosphate, monolayers on water surface by dropping method. Angewandte Chemie International Edition. 2000;**39**:1906-1920

[67] Fuhrhop JH, Wang TY. Bolaamphiphiles. Chemical Reviews. 2004;**104**:2901-2938

[68] Khalfallah A, Mazzouz S. Synthesis and determination of critical micelle concentration of betainthioates surfactants. Journal of Surfactants and Detergents. 2021;**24**:193-198

[69] Berchel M, Lemiègre L, Trepout S, Lambert O, Jeftic J. Synthesis of unsymmetrical saturated or diacetylenic cationic bolaamphiphiles. Tetrahedron Letters. 2008;**49**(52):7419-7422

[70] Mao G, Tsao Y, Tirrell M, Davis HT, Hessel V, Vanesch J, et al. Monolayers of bolaform amphiphiles: Influence of alkyl chain length and counterions. Langmuir. 1994;**10**(11):4174-4184

[71] Meister A, Blume A. Self-assembly of bipolar amphiphiles. Current Opinion in Colloid & Interface Science. 2007;**12**(3): 138-147

[72] Masuda M, Shimizu T. Multilayer structure of an unsymmetrical monolayer lipid membrane with a 'head-to-tail' interface. Chemical Communications (Cambridge, England). 2001;**23**:2442-2443

[73] Kim T, Viard M, Afonin KA, Gupta K, Popov M, Salotti J. Characterization of cationic bolaamphiphile vesicles for siRNA delivery into tumors and brain. Molecular Therapy–Nucleic Acids. 2020; **20**:359-372

[74] Fariya M, Jain A, Dhawan V, Shah S, Nagarsenker MS. Ethosuximide affects paired-pulse facilitation in somatosensory cortex of WAG\Rij Rats as a model of absence seizure. Advanced Pharmaceutical Bulletin. 2015;**5**(4): 483-491

[75] Weissig V, Lasch L, Erdos G, Meyer HW, Rowe TC, Hughes J. DQAsomes: a novel potential drug and gene delivery system made from dequalinium. Pharmaceutical Research. 1998;**15**:334-337

[76] Hutter T, Linder C, Heldman E, Grinberg S. Interfacial and self-assembly properties of bolaamphiphilic compounds derived from a multifunctional oil. Journal of Colloid and Interface Science. 2012;**365**(1):53-62

[77] Safran SA, Pincus PA, Andelman DA. Theory of spontaneous vesicle formation in surfactant mixtures. Science. 1990;**248**:354-356

[78] Kumaran V. Spontaneous formation of vesicles by weakly charged membranes. Journal of Chemical Physics. 1993;**99**:5490-5499

[79] Bergstrom M. Thermodynamics of vesicle formation from a mixture of anionic and cationic surfactants. Langmuir. 1996;**12**(10):2454-2463

[80] Grinberga S, Kolota V, Linderb C, Shaubia E, Kasyanova V, Deckelbaumc RJ, et al. Docosahexaenoic acid domains: the ultimate non-raft membrane domain. Chemistry and Physics of Lipids. 2008;**153**(2):85-97

[81] Ikeda K, Yasuda M, Ishikawa M, Esumi K, Meguro K, Binana-Limbele W, et al. Physicochemical properties of α, ω-type bolaform surfactant in aqueous solution. Eicosane-1,20-bis (triethylammonium bromide). Colloid & Polymer Science. 1989;**267**:825-830

[82] Zana R, Yiv S, Kale KM. Chemical relaxation and equilibrium studies of association in aqueous solutions of bolaform detergents. 3. Docosane-1,22-bis(trimethylammonium bromide). Journal of Colloid and Interface Science. 1980;**77**:456-465

[83] Meguro K, Ikeda K, Otsuji A, Taya M, Yasuda M, Esumi K. Physicochemical properties of the α,ω-type surfactant in aqueous solution. Journal of Colloid and Interface Science. 1987;**118**:372-378

[84] Menger FM, Wrenn S. Interfacial and micellar properties of bolaform electrolytes. Journal of Chemical Physics. 1974;**78**:1387-1390

[85] Karsa DR, Houston J. What Are Surfactants? Blackwell Publishing Ltd; 2006. pp. 1-23

[86] Boeije GM, Cano ML, Marshall SJ, Belanger SE, Compernolle VR, Dorne PB, et al. Ecotoxicity quantitative structure-activity relationships for alcohol ethoxylate mixtures based on substance-specific toxicity predictions. Ecotoxicology Environmental Safety. 2006;**64**(1):75-84

[87] Talmage S. Environmental and Human Safety of Major Surfactants: Alcohol Ethoxylates and Alkylphenol Ethoxylates. Boca Raton (FL): Lewis Publishers; 1994. p. 374

[88] De-Wall SL, Wang K, Berger DR, Watanabe S, Hernandez JC, Gokel GW. Azacrown ethers as amphiphile headgroups: Formation of stable aggregates from two- and three-armed lariat ethers. The Journal of Organic Chemistry. 1997;**62**:6784-6791

[89] Monserat K, Gratzel M, Tundo P. Light-induced charge injection in functional Crown ether vesicles. Journal of American Chemical Society. 1980;**102**: 5527-5529

[90] Djedovic NK, Ferdani R, Schlesinger PH, Gokel GW. Aggregation of lariat ethers attached to a membrane anchoring unit. Tetrahedron. 2002;**58**: 10263-10268

[91] Elbert R, Folda T, Ringsdorf H. Saturated and polymerizable amphiphiles with fluorocarbon chains. Investigation in monolayers and liposomes. Journal of American Chemical Society. 1984;**106**(25): 7687-7692

[92] Jouani MA, Szonyi S, Trabelsi H, Dieng SY, Cambon A. Synthesis and aggregation properties of new fluorine-containing double-chain amphiphiles derived from di- and tri-substituted ureas. Supramolecular Science. 1995;**2**: 117-123

[93] Riess JG. Introducing a new Element - Fluorine -Into the Liposomal Membrane. Journal of Liposome Research. 1995;**5**:413-430

[94] Krafft MP. Fluorocarbons and fluorinated amphiphiles in drug delivery and biomedical research. Advanced Drug Delivery Reviews. 2001;**47**:209-228

[95] Pedersen C. Cyclic polyethers and their complexes with metal salts. Journal of the American Chemical Society. 1967; **89**(10):2495-2496

[96] Gu X-P, Ikeda I, Okahara M, Kim L. Synthesis of 3,5-Bis(2,5,8-trioxaeicosyl)-18-crown-6 and Its Formation of Bilayer

Membranes. Chemistry Letters. 1986;**15**(10):1715-1718

[97] Shinkai S, Nakamura S, Manabe O, Yamada T, Nakashima N, Kunitake T. Synthesis and properties of membranous surfactants bearing an anion-capped crown ring as a head group. Chemistry Letters. 1986;**15**(1):49-52

[98] Khalfallah A, Boughariou B, Hedhli A. Synthesis and Characterization of Some Nonionic Bolaamphiphiles. Journal of Surfactants and Detergents. 2013;**16**: 191-196

[99] Kuo PL, Ikeda I, Okahara M. Surface properties of long chain alkyl crown ethers. Tenside Surfactants Detergents. 1982;**19**:204-206

[100] Ikeda I, Emura H, Okahara M. Hydrophilicity and complexing ability of alkyl lariat ethers. Bulletin of the Chemical Society of Japan. 1984;**57**(6): 1612-1616

[101] Muzzalupo R, Trombino S, Iemma F, Puoci F, Mesa CL, Picci N. Preparation and characterization of bolaform surfactant vesicles. Colloids Surf B, Biointerfaces. 2005;**46**:78-83

[102] Muzzalupo R, Nicoletta FP, Trombino S, Cassano R, Iemma F, Picci NA. A new crown ether as vesicular carrier for 5-fluorouracil: Synthesis, characterization and drug delivery evaluation. Colloids Surf. B, Biointerfaces. 2007;**58**:197-202

[103] Hill K, RybinskiWV SG. Alkyl Polyglycosides: Technology, Properties and Applications. Weinheim: VCH; 1997

[104] Von Rybinski W, Hill K. Alkyl Polyglycosides-properties and applications of a new class of surfactants. Angewandte Chemie (International Ed. in English). 1998;**37**:1328-1345

[105] Havlinova B, Kosik M, Kovac P, Blazej A. Synthesis and properties of surfactants based on carbohydrates. I Synthesis of some O-dodecyl aldoses. Tenside Surfactants Detergents. 1978;**15**: 72-74

[106] Pantelic I. Alkyl Polyglucosides: From Natural-Origin Surfactants to Prospective Delivery Systems. Woodhead Publishing; 2014

[107] Geetha D, Tyagi R. Tenside, Surfactants, Detergents Alkyl poly glucosides (APGs) surfactants and their properties: A Review. Vol. 492012. pp. 417-427

[108] Balzer D. Alkyl polyglucosides. In: Robb ID, editor. Specialist Surfactants. Netherlands: Springer; 1997. pp. 169-207

[109] Holmberg K. Natural surfactants. Current Opinion in Colloid & Interface Science. 2001;**6**:148-159

[110] JHP T, editor. Royal Society of Chemistry. London; 1992

[111] Jin Y, Xin R, Tong L, Du L, Li M. Combination anti-HIV therapy with the self-assemblies of an asymmetric bolaamphiphilic zidovudine/didanosine prodrug. Mol. Pharma. 2011;**8**:867-876

[112] Schmid S, Ng DYW, Mena-Osteritz E, Weil Y, Wu T, Bäuerle P. Self-assembling oligothiophene–bolaamphiphiles for loading and controlled release of doxorubicin into living cells. Chemical Communications. 2016;**52**:3235-3238

[113] Brunelle M, Polidori A, Denoyelle S, Fabiano A-S, Vuillaume PY, Laurent Lewandowski S, et al. A structure–activity investigation of hemifluorinated bifunctional bolaamphiphiles designed for gene delivery. Comptes Rendus Chimie. 2009;**12**:188-208

[114] Yu G, Li J, Yu W, Han C, Mao Z, Gao C, et al. Carbon nanotube/biocompatible bola-amphiphile supramolecular biohybrid materials: preparation and their application in bacterial cell agglutination. Advanced Materials. 2013;**25**:6373-6379

[115] Bartunek V, Junkova J, Suman J, Kolarova K, Rimpelova S, Ulbrich P, et al. Preparation of amorphous antimicrobial selenium nanoparticles stabilized by odor suppressing surfactant polysorbate 20. Materials Letters. 2015; **152**(1):207-209

[116] Siska CC, Pierini C, Lau H, Latypov R, Fesinmeyer R, Litowski J. Free Fatty acid particles in protein formulations, Part 2:Contribution of Polysorbate raw material. Journal of Pharmaceutical Sciences. 2015;**104**(2):447-456

[117] Zhou C, Qi W, Lewis E, Randolph T, Carpenter J. Reduced subvisible particle formation in lyophilized intravenous immunoglobulin formulations containing polysorbate 20. Journal of Pharmaceutical Sciences. 2016;**105**(3):2302-2309

[118] Lee WG, Lee JS, Lee JP, Shin CS, Kim MS, Park SC. Effect of surfactants on ethanol fermentation using glucose and cellulosic hydrolyzates. Biotechnology Letters. 1996;**18**(3): 299-304

[119] Vance JE, Vance DE. Phospholipid biosynthesis in eukaryotes. In: Vance DE, Vance JE, editors. Biochemistry of Lipids, Lipoproteins and Membranes. Amsterdam: Elsevier; 2008. pp. 213-244

[120] Hines JD, Fragneto G, Thomas RK, Garrett PR, Rennie GK, Rennie AR. Neutron reflection from mixtures of sodium dodecyl sulfate and dodecyl betaine adsorbed at the hydrophobic solid/aqueous interface. Journal of Colloid and Interface Science. 1997;**189**: 259-267

[121] Li Y, Xu G, Luan Y, Yuan S, Zhang Z. Studies on the interaction between tetradecyl dimethyl betaine and sodium carboxymethyl cellulose by DPD simulations. Colloids and Surfaces A: Physicochemical and Engineering Aspects. 2005;**257-258**:385-390

[122] Hazrat A, Azim K, Tauqeer A, Athanasia D, Ian F, Ibrahim O, et al. Interfacial activity and micellar morphology of an imidazolium ring containing zwitterionic surfactants. Journal of Surfactants and Detergents. 2022;**25**(3):341-350

[123] Tondo DW, Leopoldino EC, Souza BS, Micke GA, Costa ACO, Fiedler HD, et al. Synthesis of a new zwitterionic surfactant containing an imidazolium ring. Evaluating the chameleon-like behavior of zwitterionic micelles. Langmuir. 2010;**26**(20):15754-15760

Chapter 3

Review on Development and Application of Eco-Benign Surfactants

Chidi Obi and Mary-Ann N. Mgbemena

Abstract

The development of eco-benign surfactants is the foundation of ingredients utilized in the pharmaceutical, laundry, household, and personal-care cleaning products. The types of surfactants, such as anionic, cationic, nonionic, zwitterionic, properties, toxicity, and thermodynamic feasibilities, are reviewed. The properties, such as critical micelle concentration (CMC), Kraft temperature, critical packing parameter (CPP), and foaming form the basis of solubility of these surfactants in reaction mixtures. The critical micelle concentration (CMC) is the fundamental concentration at which micelle formation is observed or noticed. It helps in the thermodynamic evaluation of surfactant activities in reaction mixtures. However, the Kraft temperature, which is also referred to as the melting point of micelles, is the foundation of the solubility of surfactants. The Kraft temperature is the point of intersection between the critical micelle concentration and solubility curves. The application of eco-benign surfactants is a developing trend in pharmaceutical, laundry, households, personal care, and remediation processes.

Keywords: biosurfactant, production, properties, mechanism, application

1. Introduction

Surfactant is a special chemical substance employed in the development and production of classified products utilized in the pharmaceutical, laundry, personal-care handling, metallurgy, oil and gas, and phenomena, etc. Surfactants are described as a special class of chemical substances that lower surface tension between two liquids, gas and liquid, oil and water, and water and oil [1].

Based on their chemical composition, it may act as detergents, wetting agents, emulsifiers, demulsifiers, foaming agents, or dispersants. Typically, it is composed of organic compounds having a hydrophilic head and a hydrophobic tail (see **Figure 1**). According to Obi & Idowu [2], surfactants are described as multipurpose organic compounds with tremendous applications.

The interaction of surfactants with aqueous medium constitutes the micellar environment. This aqueous orientation generates the different forms of micelles (**Figure 2**). The hydrophilic region (head), which is polar in nature, may be classified

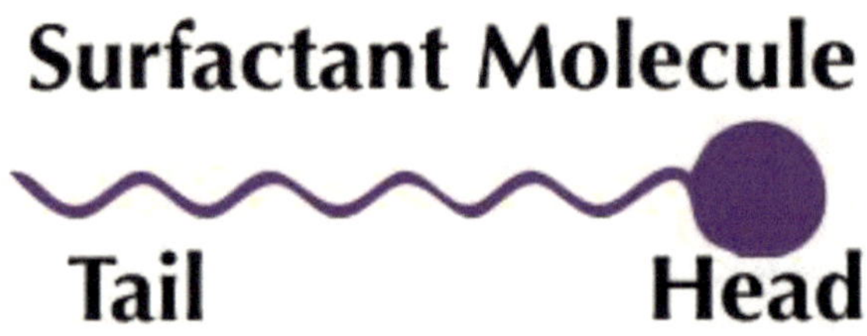

Figure 1.
Diagrammatic representation of surfactant molecule.

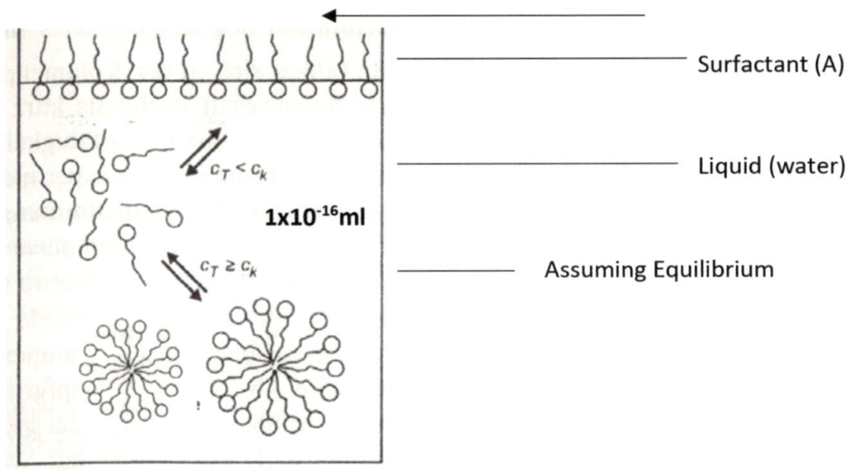

Figure 2.
Micelle formation [3].

as anionic, cationic, neutral, or zwitterionic. However, the hydrophobic region (tail), which is nonpolar, is usually compound of hydrocarbons of one or more chains with varying lengths.

In aqueous phase, these surfactants dissolve completely or with the aid of other chemical species at very low concentrations but above a certain limit called the critical micelle concentration (CMC) to form micelles (see **Figure 2**). The mechanism allows for detergency, which helps in cleaning processes [4]. The micelles formed vary in size and shape depending on the number of aggregation (N). The value of CMC can be monitored by varying the physicochemical properties of the surfactant solutions as the concentration of the surfactant molecules increases. This has been the foundation in drug formulations and management [5].

The utilization of surfactant is dated back to 1929 when the Swiss physiologist Kurt Neergaard in one of his experiments achieved a breakthrough in neonatology. The surfactant replacement therapy was actually a milestone in neonatology.

The application of surfactants is broad and varied. In the dyeing industries, that is. textiles, it helps in the penetration of dyes on the fabric [6]. Surfactants play significant role in preparation of different drug delivery systems. For drugs that are partially soluble in aqueous phase, pharmaceutically matchable surfactants (biosurfactants) or cosolvents are utilized to enhance solubility [7]. The presence of surfactants in drugs also improves their dissolution profile, permeation, and stabilization. Some examples of pharmaceutical surfactants include sodium lauryl sulfate (SLS), cetyl pyridinium chloride (CPC), sulfobetaine, and polyoxyethylene sorbitan fatty acid esters (polysorbate and tween), etc.

Surfactants are widely used as soaps and detergents in industrial, domestic, and personal-care handling activities [8]. Some of the common surfactants used in soap making include sodium stearate (SS), 4-(5-dodecyl) benzene sulfonate (DBS), dioctyl sodium sulfosuccinate (DSSA), alkyl ether phosphates (AEB), and benzalkonium chloride (BAC), etc.

Surfactants also find application in the cosmetics industries (cosmetology). Cosmetic formulations are composed of mixtures of surfactants with emulsifying, solubilizing, wetting, foaming, and dispersion properties [9]. Typical examples of surfactant used in this regard include polyethylene glycol ethers (PGE), sophorolipids, andrhamnolipids (biosurfactant), etc.

Surfactants are also referred to as oil field chemicals. They are used in oil recovery and enhancement processes and in the inhibition of corrosion during the transportation of crude oil [10]. The groups of surfactants in this category are known as emulsifiers and demulsifiers [11].

Emulsifiers and demulsifiers are amphiphilic substances. Its operations tend to promote dispersion of the phase in which they do not dissolve very well. Emulsifiers that are more soluble in water generally form oil-in-water emulsions, while those that are more soluble in oil form water-in-oil emulsions.

However, demulsifiers participate in the breaking of crude oil into oil and water phases. The operation of demulsifier is a criterion for the desirability of crude oil and the specification of pipelines. Some examples of emulsifiers include polyethylene glycol, carboxymethyl cellulose, and acacia, etc., while demulsifiers include ethoxylated phenols, ethoxylated alcohols, ethoxylated amines, and sulphonic acid salts, etc.

The application of surfactants (frothers) in mineral processing is based on contact angle. A contact angle is the angle at which hydrophobic materials form at the interphase between the liquid and the solid. The binding capacity of the mineral particles depends on wettability.

The flotation of hydrophilic or weak hydrophobic mineral particles is enhanced by the adsorption of suitable surfactants (chemical reagents) in order to lower the Gibb's free energy. Some of the examples of surfactants used as frothers (reagents) include sodium ethyl xanthate (ionizing collectors), potassium isopropyl xanthate (nonionizing collectors), and inherently hydrophobic surfactants such as kerosene, creosote, and grease, etc. However, collectors are grouped into anionic, cationic, mixed, amphoteric, bio-collectors, and ionic, etc., [12].

Recently, surfactants have been applied in the stabilization of hydrophobic nano materials in water. The successes in its application in nanotechnology have led to the development of colorimetric sensors, which have varied and massive impact on numerous interdisciplinary applications.

Surfactants are widely used in biological genetics and engineering in formulating new and in modifying existing natural surfactants, in the production of proprietary medicines, and for chemical flooding in sandstone oil reservoirs. All this is understandable since the properties of surfactants are to greater extent responsible for the existence and the performance of cellular membranes, emulsification, solubilization, and of transport of compounds that are otherwise insoluble in living tissues.

The demand for eco-benign surfactant popularly referred to as biosurfactant has generated global concerns. These groups of new generations of eco-benign surfactant molecules often directly or indirectly obtained from renewable building precursors can be broadly termed as sustainable surfactants, which are increasingly becoming popular in many application activities such as microbial enhanced oil recovery (MEOR), reduction of carbon dioxide (CO_2) emission, metals remediation, laundry industry, cosmetic industry, and food processing industry, etc.

2. Types of surfactants

There are basically two major types of surfactants based on the charge on their hydrophilic end namely the nonionic surfactants, which have neither charge nor ion attached to the hydrophilic end, and the ionic surfactants with different types of charges or ions, either positive or negative or both attached to the hydrophilic end. Ionic surfactants are further grouped into three types based on the particular type of ion attached to the hydrophilic end: anionics, cationic and amphoteric surfactants.

2.1 Nonionic surfactants

These are surfactants without any charge or ion attached to their hydrophilic end and as a result do not ionize in aqueous solutions. They are usually neutral and compatible with other types of surfactants. They are polar and their polarity is derived from the rich oxygen molecule of the hydrophilic end and the large organic molecule end of the hydrophobic end. They also possess properties such as wetting, spreading, emulsifying, solubilizing, detergency, and foaming actions. Some nonionic surfactants are non-foaming which makes them effective materials for low-foaming detergents. Some nonionic surfactants possess high ratio of hydrophilic and hydrophobic moieties and as a result, have no cloud point, while most others possess cloud point. They are used in shampoos, conditioners, and shower gel. Examples of nonionic surfactants include ethylene oxide, propylene oxide, and alcohol ethoxylates (**Figure 3**). Some natural nonionic surfactants have been synthesized from sugar-based molecules to provide safer, milder, better, and nontoxic effects, for examples include saponins such as soapwort, soap nut, and alkyl poly glucosides such as coco glucosides, decyl glucosides, and lauryl glucoside (see **Figure 4**).

Other anionic surfactants include all cleansing or foaming products such as shampoo, shower gel, foam for bathing, micellar water, and shaving cream, etc.

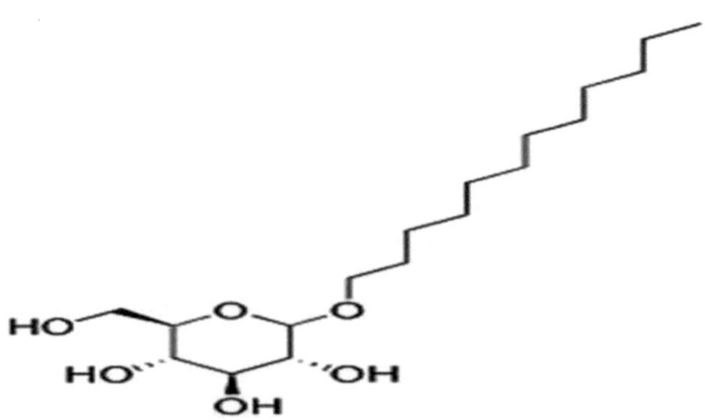

Figure 3.
Lauryl glucoside (pubchem.ncbi.nlm.nih.gov., [13]).

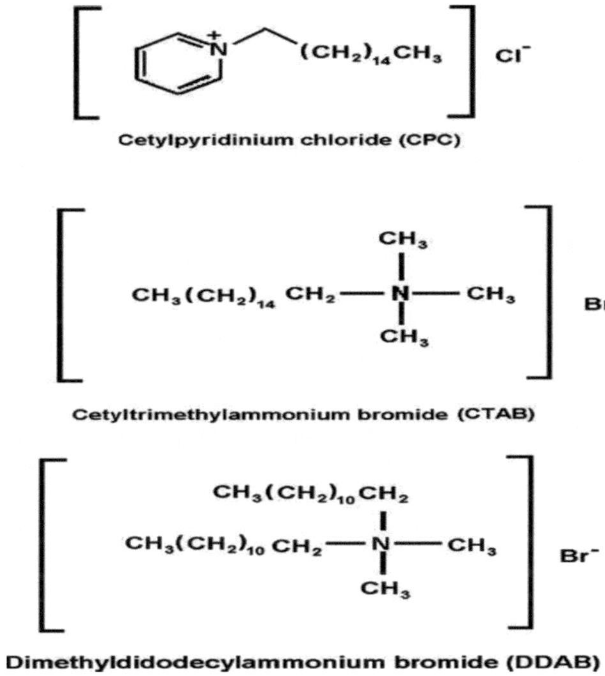

Figure 4.
Examples of cationic surfactants [14].

2.2 Cationic surfactants

Cationic surfactants are surfactants containing positively charged ions on their hydrophilic end. They are usually used where foaming is not required as they cannot produce foam on their own. They are incompatible with anionic surfactants because the opposing charges do not combine in effect, they form insoluble and ineffective compounds, they are rather compatible with the amphoteric and the nonionic surfactants. Cationic surfactants are not used as hard surface cleaners as the positive charges have the tendency to be attracted to hard surfaces.

They are used as disinfectants because of their ability to serve as antibacterial and antifungal agents, their positive charge enables them to be used as antistatic products such as fabric softeners. They are used in conditioning hair as cleansing conditioner, conditioning balm, antifreeze, and detanglers, etc. Some common examples include cetyl pyridinium chloride (CPC), cetyl trimethyl ammonium bromide (CTAB), and dimethyl dido decyl ammonium bromide (DDAB) (see **Figure 5**), etc.

2.3 Amphoteric surfactants

Amphoteric surfactants also referred as zwitterionic surfactants contain both positive and negative charges on their hydrophilic end which cancel each other by creating a zero net charge. The reaction of amphoteric surfactants depends on the pH of a given solution, if the pH of the solution is acidic, the amphoteric surfactants become positively charged and have similar behavior as cationic surfactants with low foaming, but if alkaline, they develop negative charge and behave as anionic surfactants by providing high foaming and cleansing capabilities, at neutral pH, they have similar

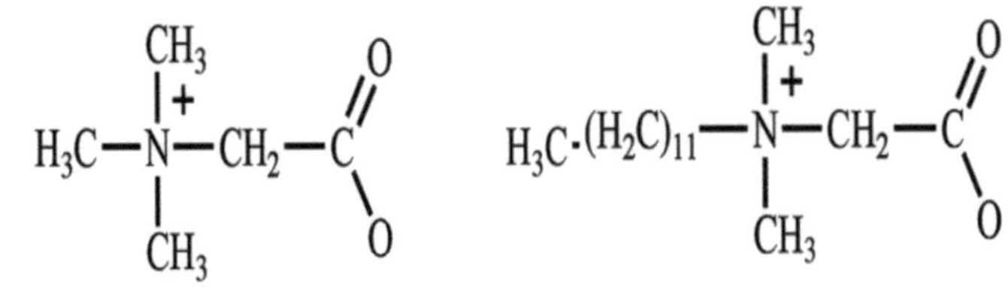

Cocamidopropyl betaine Lauryl dimethyl betaine

Figure 5.
Typical examples of amphoteric surfactants [15].

behavior as nonionic surfactants. They are compatible with all other surfactants, the anionic surfactants which they also stabilize their foam. They are widely used with nourishing and conditioning properties. Amphoteric surfactants when used alone give a gentle and milder cleansing. They are used in all cleaning, foaming, conditioning, solubilization, and antifreeze products such as cosmetics and shampoos. Common examples include amino oxides and natural ones include betaines such as cocoidopropyl betaine and amphoacetates such as sodium cocoamphoacetate and sodium lauroamphoacetate (see **Figure** 5).

3. Development of surfactants

Surfactants and synthetic detergents are the major foundation of industries in the 1930s. These two special compounds have begun to gain weight since the mid of the eighteenth century, that is. from mutton oil to green oil (petroleum sulfonic acid). Experts have predicted that the doubling of population from 2000-–2060 will bring about a corresponding increment in the development and consumption of detergent from 50 million to 120 million (a net increase of 1.4 cultures [16].

Surfactant design requires a thorough consideration of the selection of the hydrophile and hydrophobe pair such that they can be readily synthesized with minimal purification and posses optimum properties for a given application [17].

There have been massive works by researchers in changing the pathway of surfactant synthesis from chemical to greener and eco-benign route. In addition, there is a paradigm shift from the convectional pattern to a more sustainable production using renewable natural products.

The new era of eco-benign surfactants synthesis considers the stability of the ecosystem by choosing renewable natural raw materials.

The properties keenly looked at include renewability, cheapness and availability, less toxicity, and commercial feasibility with respect to kinetics and thermodynamics, etc.

There is a lot of synthesis of eco-benign surfactants in the literature such as preparation of alkyl polyglucoside (APG) using D-glucose and octanol [5] and consequent modification of natural surfactants from plants such as the modification of cashew nutshell liquid (CNSL) surfactant with triethanolamine used for chemical flooding in sandstone oil reservoirs [18].

Research team led by Professor Julian Eastoe has discovered a new magnet soap for oil spill remediation. The researchers broke this record by dissolving iron in various inert surfactant materials, composed of chloride and bromide ions similar to those found in products such as mouthwash and fabric conditioner. This ultimately resulted in a soap structure each with metallic center. This product was tested in the laboratory by covering test tube with a less dense layer of an organic solution. When a magnet was introduced from above, the soap reacted by breaking the surface tension between it and the organic solution, rising through that solution to reach the magnet.

4. Properties of convectional surfactants versus biosurfactants

The properties of surfactants to a large extent depend on the charge on the hydrophilic head (which could be positive, negative, neutral, or zwitterionic) and the ratios of the hydrophilic to lipophilic balance (HLB) [19]. The following properties are discussed in this chapter.

4.1 Critical micelle concentration (CMC) and micellar growth

This is a fundamental characteristic of surfactant development. It is the minimum concentration at which micelles are observed [20]. It is also the concentration in water at which individual molecules of surfactant begin to aggregate [21].

The aggregation formed occur within a narrow concentration range thereby causing some physicochemical properties such as surface tension, osmotic pressure, solubility, conductivity, absorbance, etc., of the surfactant solution change abruptly (see **Figure 6**).

The aggregation of surfactant molecules by way of propagation initiates micellar growth. The higher the CMC value, the larger the growth of the micelles. Importantly, a high CMC value signifies that the surfactant is hydrophilic and vice versa. At concentrations above CMC, surfactants solubilize more hydrophobic organic compounds. CMC is also a determinant factor in defining surfactants' antibacterial properties.

According to Perinelli et al. [22], surfactants with lower CMC demonstrate higher germicidal activity. Evidently noticed at concentrations above CMC, the freely dissociated surfactant molecules in the water bulk phase begin to form closed micellar

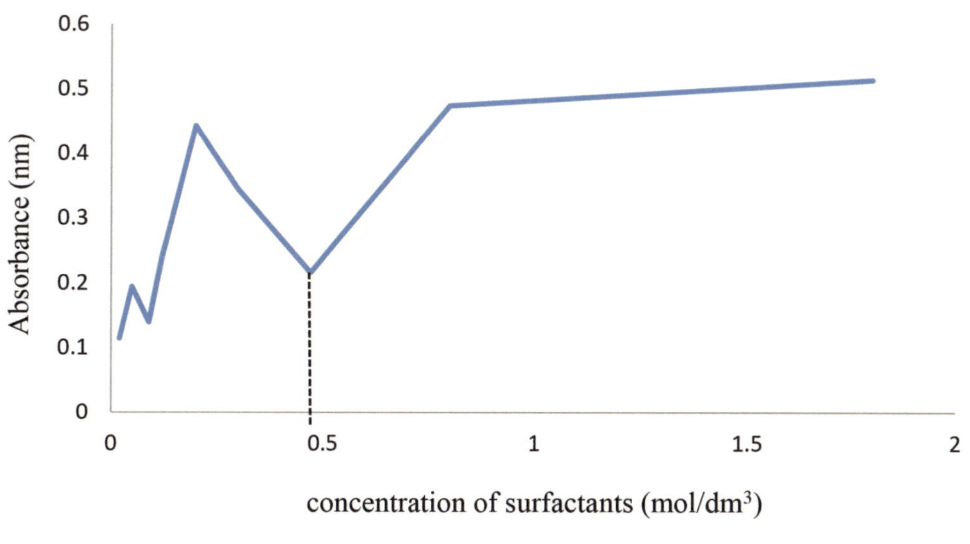

Figure 6.
Variation of absorbance with surfactant concentration [5].

structures that can be spherical, oblate, prolate, tablet, and rod-like (cylindrical) in shapes. The actual radii of these micellar structures formed are determined by surfactant concentration and geometry.

However, researches have shown that biosurfactants have low CMC and are generally effective in the lowering of surface/interfacial tension (Aranda et al., 2007; [23, 24]). It was also reported that their efficiencies can occur in extreme conditions such as temperature, pH, and salinity, etc. A group of researchers evaluated the interfacial and micellization behavior by mixing surfactin biosurfactant and synthetic surfactant sodium dodecylbenzene sulfonate (SDOBS). The result revealed that the associated molecules of surfactin and SDOBS demonstrated reduced electrostatic repulsion and promoted hydrophobic interaction between the molecules. The mole fraction of the biosurfactant in these mixed micelles was found 86% greater than its mole fraction in the bulk solution, indicating higher interfacial and self-assembly characteristics of surfactin relative to SDOBS [25–27].

4.2 Solubility and Kraft temperature

These two properties are key to making sure that surfactant molecules dispersed homogeneously on the surface or substrate. It is a chief factor in the application of surfactants in pharmaceutical and emulsion industries. Owing to the nature of the surface, that is. hydrophobic or hydrophilic, surfactants have limited solubility, which needs to reach the level of the CMC of the surfactant to be able to allow maximum performance.

The point of intersection between the solubility curve and the CMC curve of surfactants is referred to as the Kraft temperature (also called the melting point) (see **Figure** 7) [28].

This is a point that shows the ability of an insoluble surfactant to become soluble. The Kraft point increases irregularly with increased alkyl chain length. The Kraft

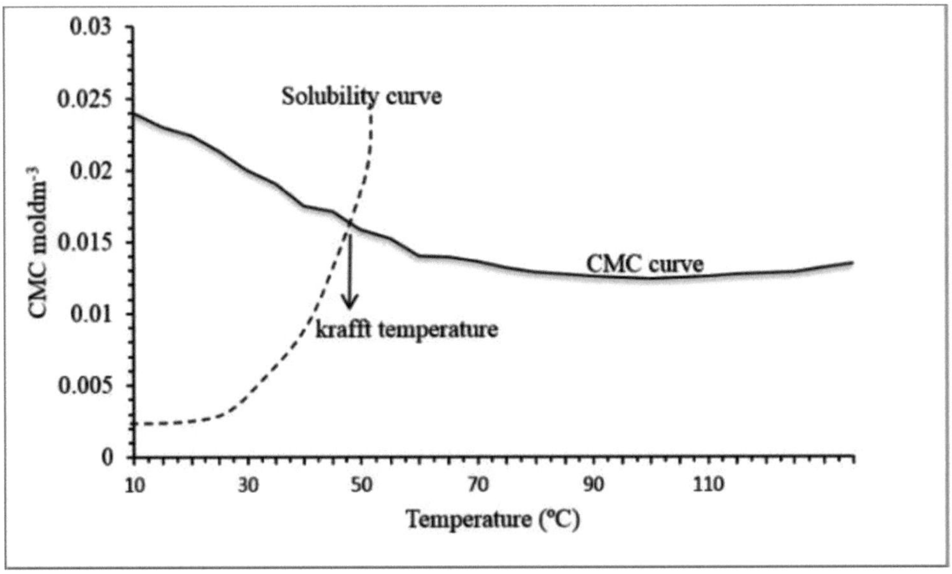

Figure 7.
Variation of CMC and solubility as a function of temperature [2].

point can also be enhanced by addition of salt or decreased by the presence of other co-solutes. The solubility is equally enhanced by the manipulation of HLB values. A low value of HLB indicates a hydrophobic surfactant, while a high HLB value indicates a hydrophilic surfactant.

Experiment has shown that the solubility of hydrophobic organic compounds in the presence of biosurfactant depends on their relative concentrations, as well as the pH and salinity of the aqueous solutions. For example, the solubility of styrene in water was enhanced 30 times (from 3.1 to 84.3 mg/dL) when the rhamnolipid concentration in solution exceeded the CMC (0.2 mg/dL) [29, 30]. Rhamnolipids, a biosurfactant, can enhance solubilization of hydrophobic compounds starting at the CMC, with solubilization increasing with increasing hydrophobicity of the micelles [31].

4.3 Foaming

This process involves the lowering of surface tension in the presence of a surfactant. The initial foam height produced by the dispersion of surfactant solution onto another identical surfactant solution per unit time is a measure of foamability. However, the decay of the foam per unit time is a measure of foam stability. The foamability and stability of foam are commonly measured using the methods of Ross-Miles [32] and Bikerman [33]. Both parameters are intrinsic in different applications.

Pradhan & Bhattacharyya [34] are proposing for eco-benign surfactants with efficient surface and foaming characteristics.

4.4 Emulsification and critical packing parameter (CPP)

Emulsification is a thermodynamic process, which involves the formation of a liquid–liquid interface. Based on the structure of surfactants, microemulsions contain both nonpolar and polar regions, allowing the numerous industrial applications. The HLB and phase inversion temperature (PIT) are the two basic parameters in determining the behavior of surfactant in emulsions. The HLB has been discussed above. However, PIT depends on the structure of the surfactant and the nature of oil. It is noted that above PIT, the emulsion is of water-in-oil type, and below PIT, the emulsion is of oil-in-water type.

Jonsson et al. [35] noted that formation of thermodynamically stable microemulsions requires that the surface tension is extremely low or else emulsion may inform instead. He further noted that water-in-oil microemulsions in equilibrium with excess oil is called a Winsor I system, an oil-in-water microemulsion in equilibrium with excess water is referred to as Winsor III system, and a microemulsion in equilibrium in both excess water and oil is called Winsor II system.

Israelachvili [36] observed that CPP is a geometrical concept that is defined as the ratio of the volume (V) of the hydrophobic tail of the surfactant to the product of the optimal areas (a) of the hydrophilic head and the length (L) of the hydrocarbon chain. It is expressed mathematically thus;

$$CPP = \frac{\mathrm{V}}{\mathrm{a} \times \mathrm{L}}$$

The CPP greatly influences the phase behavior of the surfactant. He revealed the following:

1. A low CPP ($<\frac{1}{3}$) indicates that spherical micelles can be formed (i.e., favoring oil-in-water emulsions).

2. A high CPP (>1) favors the formation of reversed micelles (i.e., favoring water-in-oil emulsions).

3. An intermediate CPP ($\frac{1}{3}-1$) will favor the formation of hexagonal, lamellar, and cubic micelles.

It is reported that hydrophobic surfactants will, thus, have high CPP values and weak hydrophobic will have lower CPP values. Addition of smaller hydrophiles to the surfactant solution has the tendency of increasing the CPP.

The CPP and PIT are somehow connected since an increase in the CPP decreases the PIT.

Extremely unfavorable packing possibilities might result in an inability of some amphiphilic substances to form micelles; such compounds have no CMCs, and not surfactants in a practical sense, but might still function as hydrotropes.

However, pieces of evidence have shown that biosurfactants as anti-agglomerants for ice, and hydrates provide improved efficiency for oil production [37]. The implication is that anti-agglomerants yield water-in-oil emulsions that limit hydrate growth to water droplets dispersed in oil [37]. The low CMC values and high emulsifying activity (formation of biofilms) (E_{24} > 70%) exhibited by biosurfactants may make them attractive for applications where low surfactant concentrations are desired [38, 39].

4.5 Detergency and dispersion stability

The ability of the surfactant molecules to solubilize with the solid (dirt) in the aqueous phase is referred to as detergency. It is a fundamental property of soaps and most detergents used in laundry, domestic, and personal-care handling processes. The spreading abilities of the aggregates of surfactant molecules are referred to as dispersion ability.

The dispersion ability of surfactant molecules is estimated in terms of amounts of particles/aggregates (N) per amount of surfactant (gram per gram) that can be dispersed.

The detergency is achieved through adsorption and desorption. Detergency occurs through the following mechanisms:

1. Formation of emulsions

2. Solubilization

3. Creation of repulsive forces between fabric and the oily dirt.

Emulsion is best performed at the PIT of the particular oil-in-water surfactant system.

However, research has also shown that biosurfactant such as lipopeptide (Iturin A) showed enhanced resistance to flocculation and creaming instabilities in emulsion compared with surfactin and fencygin [40].

4.6 Adsorption and wetting

The adsorption process encountered by surfactant molecules is easily explained by using regular solution theory and Langmuir equation. The thermodynamics and kinetics of micellization are clearly described by adsorption. The thermodynamics of surfactant systems is very important, both in theory and in practice. This is because surfactant systems represent a system that is intermediate between ordered and disordered states. The thermodynamics of micelle formation is based on the interactions arising between micellar chains within the micellar core, polar head constituents, and the surrounding medium. It is a temperature-dependent evaluation. The thermodynamic parameters associated with this type of interactions include Gibb's function of micellization, enthalpy change of micellization, and entropy change of micellization.

Obi & Idowu [2] reported that increase in temperature invokes a reduction in the hydration of the hydrophilic oxyethylene group, which in turn favors micelle formation.

The wetting process is achieved by the adsorption phenomenon. A wetting process strictly depends on properties of a surface of three phases (i.e., air, liquid, and solid). It can be enhanced by the addition of surfactant. Wetting to a large degree also depends on the physical properties of wettable solid. The equilibrium is achieved much quicker on irregular flat surface than porous surface.

The surface is characterized by high wetting, and the degree of wetting is determined by kinetics and thermodynamics of the wetting properties phenomenon. It can occur in three ways namely:

1. Spreading

2. Immersion

3. Adhesion

In spreading, a liquid placed on a solid surface increases contact surface, displacing a gas phase from a solid surface.

In adhesion, the liquid–solid contact surface does not change. The contact angle created will increase with the addition of surfactant, down to an optimum minimum of 0°, which means complete droplet spreading.

It has been established that adsorption increases as the CPP increases, although that is not always the case.

Generally, an increase in the hydrophobicity of a surfactant (biosurfactant) will increase the wetting ability of the surfactant toward a hydrophobic surface. Hence, biosurfactants use wettability property by creating a suitable environment for the attachment of bacterial adhesion thereby forming biofilms. Biofilms are the microbial communities that produce extracellular matrix [41].

5. Mechanism of biosurfactant synthesis

Biosurfactant mechanisms refer to step-by-step processes of actions, for example. Synthesis, reactions, dispersion processes, and other activities. They can be produced by enzymes, microorganisms, plants, and animals on microbial cell surfaces. Biosurfactants contain both hydrophobic and hydrophilic tail ends and act by accumulating at the interface between two immiscible liquids or between a liquid and a solid and lower their surface and interfacial tensions thereby lowering the repulsive forces between the two different phases, allowing the different phases to mix and interact more readily with themselves.

Based on the structural makeup of biosurfactants, two synthetic pathways were proposed by Sourav et al. [42]. Firstly, one pathway leads to synthesis of hydrophiles (enzymatic synthesis). However, the second pathway leads to the synthesis of hydrophobes (microbial synthesis).

The hydrophobic components are synthesized by simple lipid mechanism. However, the hydrophile components display a greater degree of structural complexity. Therefore, wide varieties of synthetic routes are adopted.

According to Syldatk and Wagner [43], four principal sequences of biosynthesis of such biosurfactant molecules (amphiphiles) can be adopted:

1. Both the hydrophilic and hydrophobic components are synthesized *de novo* by two separate pathways.

2. The hydrophilic component is synthesized *de novo,* whereas the synthesis of the hydrophobic component is induced by substrate.

3. The hydrophobic component is synthesized *de novo,* whereas the hydrophilic component is made substrate dependent.

4. The synthesis of both hydrophobic and hydrophilic components depends on the type of substrate utilized.

Sourav et al. [42] noted that the recovery of the biosurfactants after synthesis is simple, cost-effective, and eco-benign (see **Table 1**).

Biosurfactants are basically synthesized following enzymatic and microbial pathways [41]. Enzymatic synthesis can be prepared *via* biocatalysis involving the use of enzymes as catalyst. They can be enhanced to possess the desired physicochemical properties with respect to the applications. Enzymes, such as lipases, glucosidases, and phospholipases, etc., are majorly adopted in the synthesis of surfactants [44]. The various examples of enzymatic synthesis that lead to production of biosurfactants include synthesis of monoglycerides [45], synthesis of sugar esters [46], synthesis of alkyl glycosides [47, 48], and synthesis of lysophospholipids [45], etc.

S/N	Recovery technique	Biosurfactants recovered
1.	Solvent extraction	Liposan and trehalolopids, etc.
2.	Centrifugation	Trehalolipids
3.	Adsorption	Glycolipids and lipopeptides, etc.
4.	Crystallization	Glycolipids and cello biolipids
5.	Ultrafiltration	surfactin, glycolipids
6.	Foam separation and precipitation	surfactin

Table 1.
Recovery techniques and types of biosurfactants recovered [42].

However, microbial synthesis involves the use of microorganisms to produce varieties of surfactants mostly on cell surfaces or excreted extracellularly. Microbial synthesis of biosurfactants generally follows separate pathways to form the hydrophilic and hydrophobic moieties, which are then subsequently combined [43]. These two moieties may be formed through *de novo* synthesis, or they may be derived from carbon substrates available to the cells from their environment. Hence, because of the simplicity, cost-effectiveness, eco-benignity, and availability of these microbes, we are going to dwell more on their synthetic pathway.

5.1 Production of biosurfactants by microorganism

Microorganisms naturally produce microbial-derived surfactants called biosurfactants. Microorganisms produce biosurfactants due to their ability to survive on some hydrophobic substances and also the desorption processes from the hydrophobic substances, which allow direct contact with the cell, which hitherto increases the bioavailability of the insoluble substances and changes the properties of the microorganisms' cell surfaces. Microorganisms require some sources, such as carbon, nitrogen, minerals, vitamins, and growth factors, etc., to propagate and form products. These microbially derived surfactants have different chemical structures and hence diverse chemical properties. Ramkrishna et al. [44] noted that every biosurfactant has its natural role in the life cycle of the microorganism that produces it.

Biosurfactants-producing microorganisms can produce biosurfactants in aqueous medium using carbon sources such as glucose, fructose, glycerol, olive oil, agricultural waste such as sugarcane molasses and corn steep liquor [49–51], etc., and nitrogen sources such as urea or dried stalks as part of cultivation left over as substrates. In formulating the production medium, the first step required is the balanced equation based on the cell and product generation as:

Carbon and energy source + Nitrogen source + Heat + Nutrients → Carbon dioxide + Water + Biomass + Products.

Many types of biosurfactants are in existence, which includes glycolipids, phospholipids, lipopeptides, fatty acids/neutral lipids, and polymeric biosurfactants. Three examples of glycolipids include rhamnolipids, trehalolipids, and sophorolipids [52]. Rhamnolipids are produced by *Pseudomonas aeruginosa* and *P. aeruginosa*, Trehalolipids are produced by *Mycobacterium tuberculosis, Rhodpococcus erythropolis, Arthrobacter spp., Nocardia spp.* and *Corynebacterium spp*. and Sophorolipids are produced by *Torulopsis bombicola, Torulopsis petrophilum, and Torulopsis apicola*. Typical example of phospholipids includes phosphatidylethanolamine produced by *Acinetobacter spp.* and *Rhodococcus erythropolis*. Example of fatty acid biosurfactant is Spiculisporic acid produced by *Penicillium spiculisporum.* Examples of Lipopeptides biosurfactants include surfactin produced *by Bacillus subtilis* (**Figure 8**), and Lichenysin produced by *Bacillus licheniformis*. Examples of Polymeric biosurfactants include Emusan produced by *Acinetobacter cacoaceticus* RAG-1, Alasan produced by *Acinetobacter radioresistens* KA-53, Liposan produced by *Candida lipolytica,* etc. [54].

Formulating the culture media is done by introducing pure compounds essential for the growth of microorganisms with properties such as high selectivity toward the target products over the unwanted products, ability to form a persistent product, and easily disinfected, etc. In developing a suitable medium for biosurfactants, two important considerations are made firstly, it should be noted that different

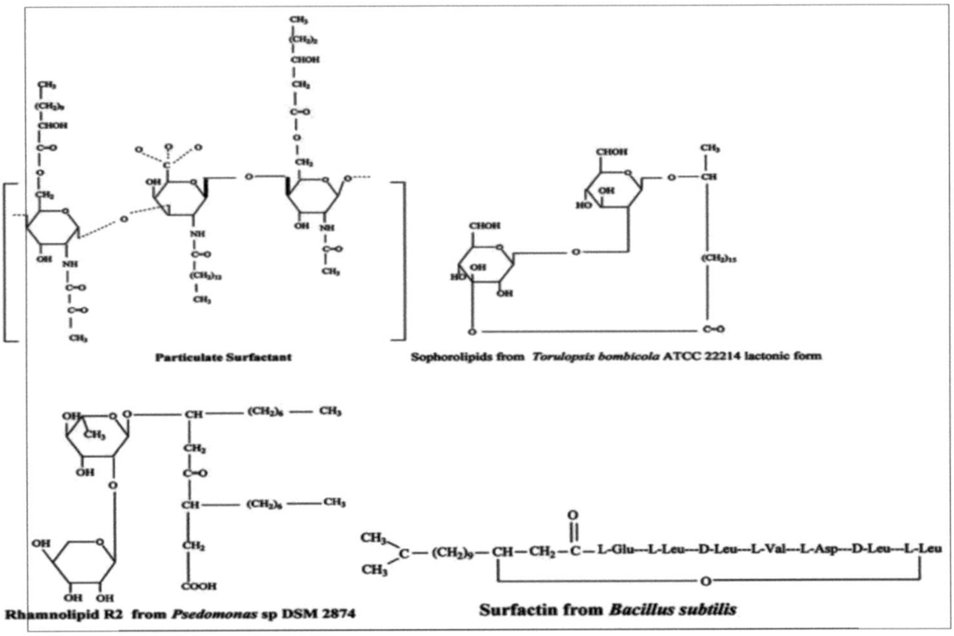

Figure 8.
Typical structures of biosurfactants produced by microbes [53].

microorganisms require different quantities of different nutrients and that each microorganism has a different metabolic pathway and also a different way of factoring the sugar and lipid moieties in its chemical structure, secondly, some microorganisms have high salt intolerance so the use of adequate quantity of chelating agents in the medium to facilitate the solubilization of the iron components to avoid the inhibition of their metabolic pathways.

To produce a hydrophilic moiety of biosurfactants for the growth of microorganisms, water-soluble substrates of carbohydrates are used, and the hydrophobic substrates, such as fats and oils, are used for building the hydrophobic moiety. Some of the pathways for the production of precursors of biosurfactants productions require carbon sources that are natural in the culture medium. The movement of carbon can be regulated *via* lipogenic pathway and glycolytic pathway restrained by microbial metabolism. A water-soluble substrate such as glucose is broken down to glucose-6-phosphate (G6P) by glycolytic pathway. In the formation of hydrophilic moieties, a variety of enzymes are used in catalyzing G6P, while in the formation of hydrophobic moiety, glucose is oxidized to pyruvate that can transform into CoA for the synthesis of malonyl-CoA react with oxaloacetate that can be converted into fatty acid for the production of lipid (**Figure 9**) [56].

To complete the process of biosurfactant formation, some multienzyme complexes are needed after the production of lipid and sugar moieties. The two most common biosurfactants are rhamnolipids produced by *P. aeruginosa* and surfactin synthesized by *B. subtilis*. Rhamnolipids production is done using two glycosyl transfer reactions, with each catalyzed with a different rhamnosyltransferase. Surfactin production is catalyzed non-ribosomally by a large multienzyme peptide synthetase complex called surfactin synthetase [57].

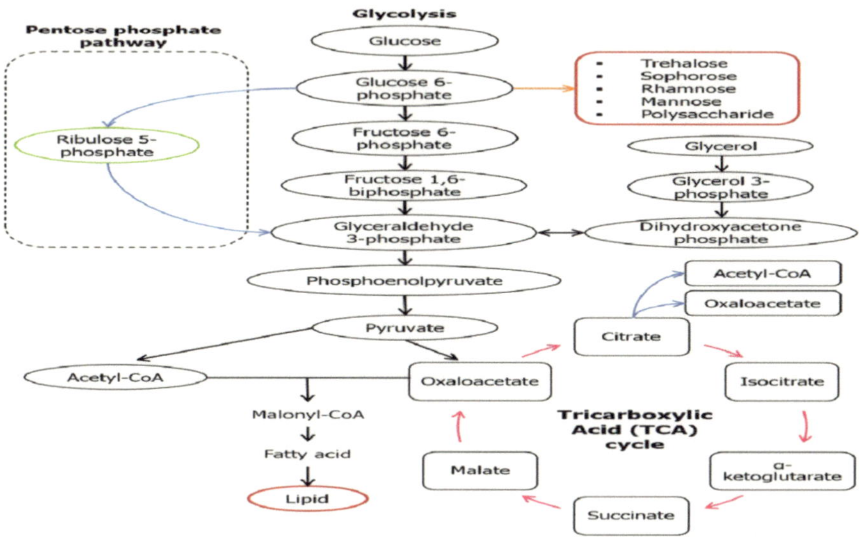

Figure 9.
Metabolic pathways of biosurfactants using hydrocarbon substrate [55].

5.2 Mechanisms of action for biosurfactants

The four mechanisms of action for biosurfactants include the following:

1. Membrane disruption

2. Interference with quorum sensing (QS)

3. Inhibition of swarming motility and biofilm formation

4. Inhibition of tissue adhesion

Membrane Disruption: Some compounds get bound to the cell membrane through electrostatic and hydrophobic interactions. Some peptides disrupt cells by forming pores in the cell membrane, which induces the thinning of the bilayer and bilayer disorder, which eventually disrupts the cell and leads to the death of cells. Example surfactin when bound to cell membrane becomes sensitive to membrane dipole potential, negatively charged lipids, and acyl chain length causing degradation of the polar region and disordering effects leading to solubilization and leakage and eventual death of cell; Iturin A when bound to the cell membrane is sensitive to Mg^{2+} and sterols, lipid phase state and acyl chain length, while Fencygin is sensitive to Ca^{2+}, pH and ionic strength, preferential electrostatic interaction with PE, lipid aggregation and intrinsic curvature all these cause dehydration of the polar region and disordering effects leading to solubilization and leakage and eventual death of the cell [58].

Interference with quorum sensing (QS): Quorum sensing is a process of cell–cell communication, which allows bacteria to share information about cell density and adjust gene expression [59]. It is also seen as a form of intercellular communication used to coordinate the physical processes and cooperative activities of bacteria at the population level [60]. It depends on the production, secretion, and detection of small diffusible autoinducers, such as acyl-homoserine lactones, auto-inducing oligo-peptides, and auto inducer 2, which increase in concentration as a function of cell density. Steps in quorum sensing range from production of small biochemical signal molecules to the passive or active release of the signal molecules to the recognition of the signal molecules by specific receptors and finally to changes in gene regulations. Using *Pseudomonas aeruginosa* the bacteria that produce rhamnolipids as example, *P. aeruginosa* virulence traits are controlled by 2-distinct lux-type quorum sensing systems termed las and rhi named after their influence on elastase and rhamnolipid production, respectively. The las system regulates the rhi system as virulence regulators. The hierarchical circuitry controls multiple virulence traits beyond elastase and rhamnolipid, including exoprotease and toxin production, motility, and biofilm formation [61]. The Las and rhi system has been seen to significantly reduce virulence animals' infections.

Inhibition of swarming, motility, and biofilm formation: Bacterial swarming motility and exopolysaccharide contribute to biofilm formation. In *P. aeruginosa*, Psl (a repeating pentasaccharide) and Pel (a repeating pentasaccharide) are two major exopolysaccharides that serve as biofilm matrix components. Psl contains L-rhamnose, which is an essential matrix component used by *P. aeruginosa* to initiate biofilm formation and the maintenance of biofilm structure [52]. Loss of Psl production greatly attenuates *P. aeruginosa* biofilm formation [62, 63].

Inhibition of tissue adhesion: Adhesion mechanisms involve appendages capable of binding to some chemical species on certain surfaces [64].

According to Karlapudi et al. [41], microbes that produce biosurfactant undergo physiological metamorphosis as represented in **Figure 10**.

5.3 Emerging application of biosurfactant in oil spill remediation

Oil spillage is a natural or artificial phenomenon that pollutes the fragile ecosystem. In fact, it is a devastating effect to aquatic domain.

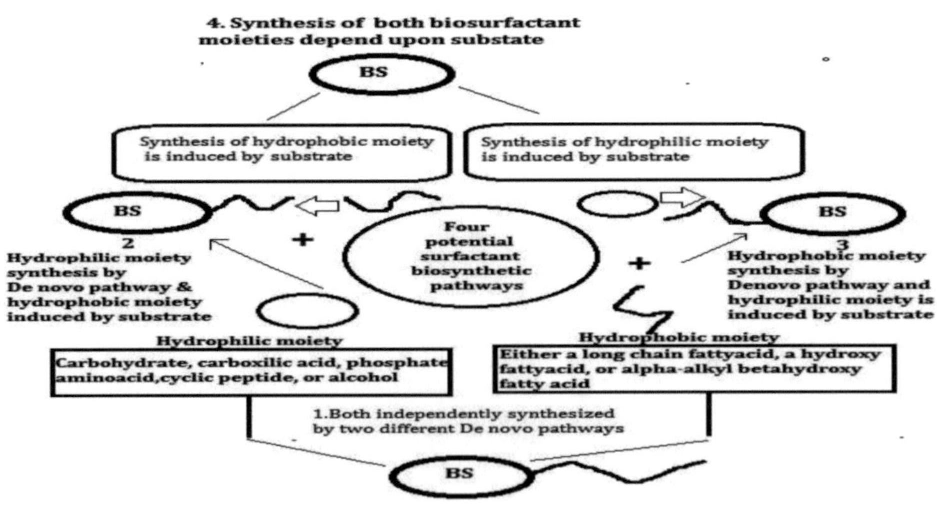

Figure 10.
Biosynthetic pathway of bacteria producing biosurfactant [13].

Surfactants, which are the key to the dispersant's effectiveness and efficiency, are compounds comprising of two different kinds of chemical groups: one that is oil-compatible (lipophilic or hydrophobic) and one that is water-compatible (hydrophilic). The lipophilic end of the surfactant molecule has London dispersion forces similar to those of the oil, allowing this portion of the molecule to be soluble in the oil layer. On the other hand, the hydrophilic end has dipole–dipole forces and/or hydrogen bonding similar to that of water, allowing this portion of the molecule to be soluble in the water. This dual action helps in the lowering of the interfacial tension between the oil and the water, allowing the oil to break up into small droplets surrounded by surfactant molecules. The selection of surfactants with the appropriate hydrophilic–lipophilic balance (HLB) values is determinant in the proper remediation of spilled oil.

The use of chemical surfactants (dispersants) had been reported for their toxicity to aquatic domains, so were, referred to as unsuitable for oil spill remediation.

However, biosurfactants are currently used in enhanced oil recovery such as microbial enhanced oil recovery (MEOR) [41, 65, 66] and chemical flooding in sandstone oil reservoirs [18].

MEOR activates the stimulation of native reservoir microbes (hydrocarbonoclastic bacteria) or mechanically injecting specially selected natural bacteria into the oil reservoir to produce specific metabolic events promoting the synthesis of biosurfactants, which ultimately lead to improved oil recovery [67–69]. This is achieved through the reduction of oil viscosity and interfacial tension, which results from a partial breakdown of the large molecular structure of crude oil, making it more fluid-like. This process leads to the production of CO_2 gas as a byproduct of microbial metabolism both pressurize the reservoir and move upward, displacing oil in the well. Additionally, the production of biomass that accumulates between the oil and the rock surface of the well displaces the oil, making it easier to recover from the well [68].

It is also applied in the recovery of residual oil with low permeability or high viscosity [67].

According to Bach et al. [70], biosurfactant adsorbs the oil by altering the wettability capacity of the porous media (spilled site). They noted that emulsion produced by *Acinetobacter venetianus* (ATCC 31012) at 0.1 mg/mL removes 89% of crude oil, which had been reabsorbed to the samples of limestone, and 98% of removal was achieved at 0.5 mg/mL.

Yakimov et al. [71] also noted that many bacterial species that produce biosurfactants had been described for the microbially enhanced oil recovery *in situ* applications that belong to *Bacillus*sps. because of their thermal and halotolerance ability. A typical *Bacillus* strain was grown and produced lichenysin by both anaerobic and aerobic processes at relatively high temperatures ranging from 40 to 60°C.

6. Challenges and way forward

The simultaneous manipulation of surfactant's structure with respect to functionality, economics, and eco-benignity possess great difficulty in the replacement of convectional surfactants (i.e., petroleum source). Surfactants used as soap are obviously sustainable under appropriate condition but show difficulty at lower temperatures and insensitivity toward hard water. This weakness is a fundamental driving force in the development of synthetic and eco-benign surfactants. The solubility of surfactants in aqueous phase also poses great challenge in its applications. The toxicity of convectional surfactants has posed serious danger to their applications. The toxicity of these surfactants is based on their structural conformation [72]. The elevated concentrations of surfactants and their degraded products interfere with microbial dynamics, and their important biogeochemical phenomena hinder plant-surviving processes and their ecological habitant, and retard the human organic and systemic functionalities [73]. The following are few toxicities X-rayed in this discussion.

Skin infection toxicity: Research has shown that ionic surfactants are the most toxic if they are soluble in water [74]. They assert that crystalline ionic surfactants of low solubility show low toxicity. However, for nonionic surfactants, the nature of the chemical bonding linking the polar head region, and the alkyl chain hydrocarbon has great impact on skin toxicity.

Aquatic perturbation toxicity: Surfactants are new emerging contaminants (EC), about 15 million tons of surfactants produced globally are routed directly into waste-water streams [75]. Aquatic toxicity varies strongly among living organisms such as fish, daphnia, algae, or bacteria. Above listed, bacteria appear to be the least sensitive species in case of acute and chronic toxicity. Examples of such surfactants that affect aquatic habitant include branched alkylbenzene sulfonates (ABS), ditallow dimethyl ammonium chloride (DTDMAC), and nonionic nonylphenol ethoxylates (NPEO), etc.

Fire-control chemicals toxicity: These are chemicals used to protect forest resources from wildland fires. These chemicals contain several components including foam. The foam concentrates are typically composed of anionic surfactants, stabilizers, and solvents. These chemicals in large concentrations and applications affect both aquatic and human ecosystems [76]. Examples of the surfactant component of the foam include sodium dodecyl sulfonates (SDS) and linear alkyl benzene sulfonates (LABS), etc.

Cellular toxicity: This occurs as a result of surfactants used as herbicides or pesticides. The composition of the herbicide (i.e., surfactants, solvent, and anti-freeze) affect the membrane activity, metabolic activity, mitochondrial activity, and total protein synthesis rate in a cell culture [77]. Research has shown that sodium lauryl ether sulfate (SLES) and polyoxyethylene lauryl ether significantly destroyed the cellular activities.

However, literature has shown that biosurfactants are generally low in toxicity [41]. In addition, several studies have explored biosurfactant toxicity in plants, aquatic life, and human cell lines. For example, Santos et al. [78] reported that biosurfactants produced from *C. lipolytica* were found to have no impact on seed germination or root length of various plant species at concentrations ranging from 0.5 to 2 times the CMC. Santos et al. [78] also reported that brine shrimp also showed no signs of cytotoxicity when exposed to a glycolipid biosurfactant.

This study encourages researchers to rise up to the challenges by propounding seamless and greener route for the development/synthesis of biosurfactants through modification of sustainable hydrophilic head groups and hydrophobic tail groups and designing other natural moieties that will be compatible to extreme conditions such as temperature, pH, and salinity, etc. [34, 79, 80]. It is reported that the utilization of low-cost microbial substrates through genetic engineering combined with improvements in the downstream purification could significantly scale up the economics, eco-benignity, and the availability of biosurfactants for large scale applications [42].

7. Conclusion

This chapter has in a nutshell evaluated the synthesis of biosurfactants which is the way forward and exposed in general the properties, applications, toxicities, and challenges of the old and new trends. It is believed that the adoption of eco-benign route of synthesis will be a panacea in solving several toxicities experienced by both human and aquatic habitants.

Therefore, this study encourages researchers to develop toxicity testing procedures and cost-effective pathways of production and recovery that will not limit its use.

Declaration of interest

The authors state no conflicts of interest concerning the preparation of this manuscript.

Author details

Chidi Obi[1]* and Mary-Ann N. Mgbemena[2]

1 Faculty of Science, Department of Pure and Industrial Chemistry, University of Port Harcourt, Nigeria

2 Department of Chemistry, Federal University of Agriculture, Umuahia, Nigeria

*Address all correspondence to: chidi.obi@uniport.edu.ng

References

[1] Hosoi J, Koyama J, Ozawa T. Chapter 5-new aspects of cosmetics and cosmetic science. In: Sakamoto K, Lochhead RY, Maibach HI, Yamashita Y, editors. Cosmetic Science and Technology. Amsterdam, Netherlands: Elsevier; 2017. pp. 87-100. DOI: 10.1016/B978-0-12-802005-0.00005-7

[2] Obi C, Idowu VA. Determination of critical micelle concentration and thermodynamic evaluations of Micellization of GMS. Modern Chemistry & Applications. 2018;**6**(2): 1-5. DOI: 10.4172/2329-6798.1000251

[3] Phillips JN. The energetics of micelle formation. Transactions of the Faraday Society. 1955;**51**:561-569. DOI: 10.1039/TF9555100561

[4] Hubbe MA, Szlek DB, Vera RE. Detergency mechanisms and cellulosic surfaces: A review. BioResources. 2022;**17**(4):7167-7249. DOI: 10.15376/biores.17.4.hubbe

[5] Obi C, Aliba PV, Iwuoha GN. Synthesis and characterization of alkyl polyglucoside surfactant: A pilot tool to industrial applications. Chemistry International. 2021;7(3):172-180

[6] Park JY, Hirata Y, Hamada K. Interactions between dyes and surfactants in inkjet ink used for textiles. Journal of Oleo Science. 2011;**60**(12): 627-637. DOI: 10.5650/jos.60.627

[7] Thakur S, Singh A, Sharma R, Aurora R, Jain SK. Biosurfactants as a novel additive in pharmaceutical formulations: Current trends and future implications. Current Drug Metabolism. 2020;**21**(11):885-901. DOI: 10.2174/1389200221666201008143238

[8] Babak B, Saied S, Hasan R. Detergents and surfactants: A brief review. Open Access Journal of Science. 2019;**3**(3): 94-99. DOI: 10.15406/oajs.2019.03.00138

[9] Lourith N, Kanlayavattanakul M. Natural surfactants used in cosmetics: Glycolipids. International Journal of Cosmetic Science. 2009;**31**:255-261. DOI: 10.1111/j.1468-2494.2009.00493.x

[10] Ahmed F, El-S R, Areej Al B, Moataz M, Ismail A, Khalid A. Assessment of new designed surfactants as eco-friendly inhibitors for the corrosion of steel in acidic environment and evaluation of their biological and surface features: Thermodynamic, kinetic and mechanistic aspects. Journal of Adhesion Science and Technology. 2022;**36**(18):1993-2019. DOI: 10.1080/01694243.2021.1997039

[11] Farzand R, Hossien E. Microbially-derived biosurfactants for improved sustainability in industry. In: Green Sustainable Process for Chemical and Environmental Engineering and Science. Elsevier; 2021. pp. 399-422

[12] Pattanaik A, Venugopal R. Role of surfactants in mineral processing: An overview. In: Surfactants and Detergents. London, UK: IntechOpen; 2019. DOI: 10.5772/intechopen.85947

[13] Fiume MM, Heldreth B, Bergfeld WF, et al. Safety Assessment of Decyl Glucoside and Other Alkyl Glucosides as Used in Cosmetics. International Journal of Toxicology. 2013;**32**(5_suppl):22S-48S. DOI: 10.1177/1091581813497764

[14] Rui AG, Krister H, Björn L. Cationic surfactants: A review. Journal of Molecular Liquids. 2023;**375**:121335. DOI: 10.1016/j.molliq.2023.121335

[15] Stephanie KC, Neil WB. Chapter 14 - betaine amphoteric surfactants—Synthesis, properties, and applications. In: Douglas GH, Daniel KYS, Richard DA, editors. Biobased Surfactants. Second ed. USA: AOCS Press; 2019. pp. 447-469. DOI: 10.1016/B978-0-12-812705-6.00014-9

[16] Randal MH. Silicone surfactants-new developments. Current Opinion in Colloid & Interface Science. 2002;**7**(5-6):255-261. DOI: 10.1016/S1359-0294(02)00068-7

[17] Athumani O, Ruibo C, Zhuoyan Z, Xingguang X. A comprehensive review of recent advances on surfactant architectures and their applications for unconventional reservoirs. Journal of Petroleum Science and Engineering. 2021;**206**:109025. DOI: 10.1016/j.petrol.2021.109025

[18] Obuebite AA, Eke WI, Udoh T. Evaluation of modified cashew nutshell liquid as natural surfactants for chemical flooding in sandstone oil reservoirs. Journal of Engineering and Applied Science. 2022;**69**:114. DOI: 10.1186/s44147-022-00167-4

[19] Saxena N, Mandal A. Surfactants and Their Types. In: Natural Surfactants. Springer Briefs in Applied Sciences and Technology. Cham: Springer; 2022. DOI: 10.1007/978-3-030-78548-2_2

[20] Perez-Rodriguez M, Prieto G, Rega C. A comparative study of the determination of the critical micelle concentration by conductivity and dielectric constant measurements. Langmuir. 2017;**14**:4422-4426. DOI: 10.1021/la980296a

[21] Nesmerak K, Nemcova I. Determination of critical micelle concentration by electrochemical means. Analytical Letters. 2019;**39**:1023-1040. DOI: 10.1080/00032710600620302

[22] Perinelli DR, Cespi M, Lorusso N, Palmieri GF, Bonacucina G, Blasi P. Surfactant self-assembling and critical micelle concentration: One approach fit all? Langmuir. 2020;**36**(21):5745-5753. DOI: 10.1021/acs.langmuir.0c00420

[23] Manko D, Zdziennicka A, Janczuk B. Thermodynamic properties of Rhamnolipid Micellization and adsorption. Colloids and Surfaces B-Biointerfaces. 2014;**119**:22-29. DOI: 10.1016/j.colsurfb.2014.04.020

[24] Nitschke M, Costa S, Contiero J. Rhamnolipids and PHAs: Recent reports on pseudomonas-derived molecules of increasing industrial interest. Process Biochemistry. 2011;**46**(3):621-630. DOI: 10.1016/j.procbio.2010.12.012

[25] Onaizi SA, Nasser MS, Twaiq FA. Micellization and interfacial behavior of a synthetic surfactant-biosurfactant mixture. Colloids and Surfaces A-Physicochemical and Engineering Aspects. 2012;**415**:388-393. DOI: 10.1016/j.colsurfa.2012.09.014

[26] Pereira JFB, Gudina EJ, Costa R, Vitorino R, Teixeira JA, Coutinho JAP, et al. Optimization and characterization of biosurfactant production by bacillus subtilis isolates towards microbial enhanced oil recovery applications. Fuel. 2013;**111**:259-268. DOI: 10.1016/j.fuel.2013.04.040

[27] Penfold J, Chen M, Thomas RK, Dong C, Smyth TJ, Perfumo A, et al. Solution self-assembly of the Sophorolipids biosurfactant and its mixture with anionic surfactant sodium dodecyl benzene sulfonate. Langmuir. 2011;**27**(14):8867-8877. DOI: 10.1021/la201661y

[28] Fakiha E-TH, Ayman EE. Gemini surfactants as corrosion inhibitors for carbon steel. Journal of Molecular

Liquids. 2017;**230**:395-407. DOI: 10.1016/j.molliq.2017.01.047

[29] Konishi M, Nagahama T, Fukuoka T, Morita T, Imura T, Kitamoto D, et al. Yeast extract stimulates production of glycolipid biosurfactants, Mannosylerythritol lipids, by Pseudozyma Hubeiensis SY62. Journal of Bioscience and Bioengineering. 2011;**111**(6):702-705. DOI: 10.1016/j.jbiosc.2011.02.004

[30] Zdziennicka A, Zdziennicka A, Janczuk B, Janczuk B. Wetting and adhesion properties of Rhamnolipid and surfactin. International Journal of Adhesion and Adhesives. 2018;**84**:275-282. DOI: 10.1016/j.ijadhadh.2018.04.005

[31] Yang Z, Cui J, Yin B. Solubilization of nitrogen heterocyclic compounds using different surfactants. Water Air and Soil Pollution. 2018;**229**(9):304. DOI: 10.1007/s11270-018-3917 – 8

[32] Ross-Miles, (2001). Standard Test Method for foaming properties of surface-active agents (ASTM D 1173-63).

[33] Bikerman JJ. General. Foam Films. In: Foams. Berlin, Heidelberg: Springer; 1973. p. 10. DOI: 10.1007/978-3-642-86734-7_1

[34] Pradhan A, Bhattacharyya A. Quest for an eco-friendly alternative surfactant: Surface and foam characteristics of natural surfactants. Journal of Cleaner Production. 2017;**150**:127-134. DOI: 10.1016/j.jclepro.2017.03.013

[35] Jonsson B, Lindman B, Holmberg K, Kronberg B. Surfactants and Polymers in Aqueous Solution. Chichester: John Wiley & Sons; 2001

[36] Israelachvili JN. Intermolecular and Surface Forces. New York: Academic Press; 1985. pp. 249-253

[37] York JD, Firoozabadi A. Comparing effectiveness of Rhamnolipid biosurfactant with a quaternary ammonium salt surfactant for hydrate anti-agglomeration. Journal of Physical Chemistry B. 2008;**112**(3):845-851. DOI: 10.1021/jp077271h

[38] Lovaglio RB, dos Santos FJ, Jafelicci M, Contiero J. Rhamnolipid emulsifying activity and emulsion stability: pH rules. Colloids and Surfaces B-Biointerfaces. 2011;**85**(2):301-305. DOI: 10.1016/j.colsurfb.2011.03.001

[39] Vaz DA, Gudina EJ, Alameda EJ, Teixeira JA, Rodrigues LR. Performance of a biosurfactant produced by a bacillus subtilis strain isolated from crude oil samples as compared to commercial chemical surfactants. Colloids and Surfaces B-Biointerfaces. 2012;**89**:167-174. DOI: 10.1016/j.colsurfb.2011.09.009

[40] Deleu M, Razafindralambo H, Popineau Y, Jacques P, Thonart P, Paquot M. Interfacial and emulsifying properties of Lipopeptides from bacillus subtilis. Colloids and Surfaces A-Physicochemical and Engineering Aspects. 1999;**152**(1-2):3-10. DOI: 10.1016/s0927-7757(98)00627 – x

[41] Karlapudi AP, Venkateswarulu TC, Tammineedi J, Kanumuri L, Ravuru BK, Dirisala VR, et al. Role of biosurfactants in bioremediation of oil pollution-a review. Petroleum. 2018;**4**(3):241-249. DOI: 10.1016/j.petlm.2018.03.007

[42] Sourav D, Susanta M, Aniruddha G, Rumpa S, Bidyut S. A review on natural surfactants. Royal Society of Chemistry Advances. 2015;**5**(81):65757-65767. DOI: 10.1039/C5RA11101C

[43] Syldatk C, Wagner F. Production of biosurfactants. In: Kosaric N, Cairns WL, Gray NCC, editors. Biosurfactants and Biotechnology. Vol. 25. New York: Marcel Dekker; 1987. pp. 89-120

[44] Ramkrishna S, Ackmez M, Gunaseelan D. Biosurfactants: Synthesis, properties and applications in environmental bioremediation. In: Mohee R, Mudhoo A, editors. Bioremediation and Sustainability. Massachusetts, USA: O Scrivener Publishing LLC; 2012. pp. 137-212. DOI: 10.1002/9781118371220.ch4

[45] Zinjarde SS, Ghosh M. Production of surface-active compounds by biocatalyst technology. In: Sen RK, editor. Biosurfactants. USA: Landes, Biosciences; 2010. pp. 289-303

[46] Bilyk A, Bistline RG, Haas MJ, et al. Lipase-catalyzed triglyceride hydrolysis in organic solvent. Journal of the American Oil Chemists' Society. 1991;**68**:320-323. DOI: 10.1007/BF02657685

[47] Hughes FA, Lew BW. Physical and functional properties of some higher alkyl polyglucosides. Journal of American Oil Chemists' Society. 1970;**47**(5):162-167. DOI: 10.1007/BF02638743

[48] Hui-Lei Y, Jian-He X, Wen-Ya L, Guo-Qiang L. Identification, purification and characterization of β-glucosidase from apple seed as a novel catalyst for synthesis of O-glucosides. Enzyme and Microbial Technology. 2007;**40**(2):354-361. DOI: 10.1016/j.enzmictec.2006.05.004

[49] Desai JD, Banat IM. Microbial production of surfactants and their commercial potential. Microbiology and Molecular Biology Reviews. 1997;**61**(1):47-64. DOI: 10.1128/mmbr.61.1.47-64.1997

[50] Gudiña EJ, Fernandes EC, Rodrigues AI, Teixeira JA, Rodrigues LR. Biosurfactant production by Bacillus subtilis using corn steep liquor as culture medium. Frontiers in Microbiology. 2015;**6**:59. DOI: 10.3389/fmicb.2015.00059

[51] Takahashi M, Morita T, Wada K, Hirose N, Fukuoka T, Imura T, et al. Production of Sophorolipids glycolipid surfactants from sugarcane molasses using Starmerella bombicola NBRC 10243. Journal of Oleo Science. 2011;**60**(5):267-273

[52] Aranda FJ, Espuny MJ, Marques A, Teruel JA, Manresa A, Ortiz A. Thermodynamics of the interaction of a Dirhamnolipid biosurfactant secreted by pseudomonas aeruginosa with phospholipid membranes. Langmuir. 2021;**23**(5):2700-2705. DOI: 10.1021/la061464z

[53] Swaranjit SC, Randhir SM, Jasminder K, Mehta SK. Synthesis of biosurfactants and their advantages to microorganisms and mankind. In: Ramkrishna S, editor. Biosurfactants. New York, USA: Landes Bioscience and Springer Science + Business Media; 2010

[54] Magdalena P, Grazyna AP, Zofia P, Swaranjit SC. Environmental applications of biosurfactants: Recent advances. International Journal of Molecular Sciences. 2011;**12**:633-654

[55] Kalyani K, Chanchal KM. Modelling the Krebs cycle and oxidative phosphorylation. Journal of Biomolecular Structure and Dynamics. 2014;**32**(2):242-256. DOI: 10.1080/07391102.2012.762723

[56] Hommel R, Huse K. Regulation of soporose lipid production by Candida (Torulopsis) apicola. Biotechnology Letters. 1993;**15**:853-858. DOI: 10.1007/BF00180154

[57] Das P, Mukherjee S, Sen R. Genetic regulations of the biosynthesis of microbial surfactants: An overview. Biotechnology & Genetic Engineering

Reviews. 2008;**25**:165-185. DOI: 10.5661/bger-25-165

[58] Balleza D, Alessandrini A, Beltrán GMJ. Role of lipid composition, physicochemical interactions, and membrane mechanics in the molecular actions of microbial cyclic Lipopeptides. Journal of Membrane Biology. 2019;**252**(2-3):131-157. DOI: 10.1007/s00232-019-00067-4

[59] Costi DS. Quorum Sensing: Bacteria Talk Sense. Clinical Infectious Diseases. 2008;**47**(8):1070-1076. DOI: 10.1086/592072

[60] Zhang W, Li C. Exploiting quorum sensing interfering strategies in gram-negative bacteria for the enhancement of environmental applications. Frontiers in Microbiology. 2016;**6**:1535. DOI: 10.3389/fmicb.2015.01535

[61] Van Delden C, Iglewski BH. Cell-to-cell signaling and Pseudomonas aeruginosa infections. Emerging Infectious Diseases. 1998;**4**(4):551-560. DOI: 10.3201/eid0404.980405

[62] Colvin KM, Irie Y, Tart CS, Urbano R, Whitney JC, Ryder C, et al. The Pel and Psl polysaccharides provide Pseudomonas aeruginosa structural redundancy within the biofilm matrix. Journal of Environmental Microbiology. 2012;**14**(8):1913-1928. DOI: 10.1111/j.1462-2920.2011.02657. x

[63] Ma L, Jackson KD, Landry RM, Parsek MR, Wozniak DJ. Analysis of Pseudomonas aeruginosa conditional Psl variants reveals roles for the Psl polysaccharide in adhesion and maintaining biofilm structure post attachment. Journal of Bacteriology. 2006;**188**(23):8213-8221. DOI: 10.1128/JB.01202-06

[64] Simone K, Andréa CDR. Bacterial adhesion to biomaterials: What regulates this attachment? A review. Japanese Dental Science Review. 2021;**57**:85-96. DOI: 10.1016/j.jdsr.2021.05.003

[65] Clements T, Ndlovu T, Khan S, Khan W. Biosurfactants produced by Serratia species: Classification, biosynthesis, production and application. Applied Microbiology and Biotechnology. 2019;**103**(2):589-602. DOI: 10.1007/s00253-018-9520 – 5

[66] Patel S, Homaei A, Patil S, Daverey A. Microbial biosurfactants for oil spill remediation: Pitfalls and potentials. Applied Microbiology and Biotechnology. 2019;**103**(1):27-37. DOI: 10.1007/s00253-018-9434-2

[67] Gao P, Li G, Li Y, Li Y, Tian H, Wang Y, et al. An exogenous surfactant-producing bacillus subtilis facilitates indigenous microbial enhanced oil recovery. Frontiers in Microbiology. 2016;7:186. DOI: 10.3389/fmicb.2016.00186

[68] Marchant R, Banat IM. Microbial biosurfactants: Challenges and opportunities for future exploitation. Trends in Biotechnology. 2012;**30**(11):558-565. DOI: 10.1016/j.tibtech.2012.07.003

[69] Safdel M, Anbaz MA, Daryasafar A, Jamialahmadi M. Microbial enhanced oil recovery, a critical review on worldwide implemented field trials in different countries. Renewable & Sustainable Energy Reviews. 2017;**74**:159-172. DOI: 10.1016/j.rser.2017.02.045

[70] Bach H, Berdichevsky Y, Gutnick D. An exocellular protein from the oil degrading. Microbe Acinetobacter venetianus RAG-1 enhances the emulsifying activity of the polymeric bioemulsifier emulsan. Applied and Environmental Microbiology. 2003;**69**(5):2608-2615

[71] Yakimov MM, Timmis KN, Wray V, Fredrickson HL. Characterization of a new lipopeptide surfactant produced by thermotolerant and halotolerant subsurface bacillus licheniformis BAS50. Applied and Environmental Microbiology. 1995;**61**(5):1706-1713

[72] Lechuga M, Fernández-Serrano M, Jurado E, Núñez-Olea J, Ríos F. Acute toxicity of anionic and non-ionic surfactants to aquatic organisms. Ecotoxicology and Environmental Safety. 2016;**125**:1-8. DOI: 10.1016/j.ecoenv.2015.11.027

[73] Badmus SO, Amusa HK, Oyehan TA, Saleh TA. Environmental risks and toxicity of surfactants: Overview of analysis, assessment, and remediation techniques. Environment and Science Pollution Research International. 2021;**28**(44):62085-62104. DOI: 10.1007/s11356-021-16483-w

[74] Lémery E, Briançon S, Chevalier Y, Bordes C, Oddos T, Gohier A, et al. Skin toxicity of surfactants: Structure/toxicity relationships. Colloids and Surfaces A: Physicochemical and Engineering Aspects. 2015;**469**:166-179. DOI: 10.1016/j.colsurfa.2015.01.019

[75] Abreham TB, Dawit NB, Ravi N, Sreenivasulu C. Recent advances in surfactant-enhanced in-situ chemical oxidation for the remediation of non-aqueous phase liquid contaminated soils and aquifers. Environmental Technology & Innovation. 2018;**9**:303-322. DOI: 10.1016/j.eti.2017.08.004

[76] Kevin JB, Steven JH. Acute toxicity of fire-control chemicals, nitrogenous chemicals, and surfactants to rainbow trout. Transactions of the American Fisheries Society. 2000;**129**:408-418. DOI: 10.1577/1548-8659

[77] Song HY, Kim YH, Seok SJ, Gil HW, Yang JO, Lee EY, et al. Cellular toxicity of surfactants used as herbicide additives. Journal of Korean Medical Science. 2012;**27**(1):3-9. DOI: 10.3346/jkms.2012.27.1.3

[78] Santos DKF, Meira HM, Rufino RD, Luna JM, Sarubbo LA. Biosurfactant production from Candida lipolytica in bioreactor and evaluation of its toxicity for application as a bioremediation agent. Process Biochemistry. 2017;**54**:20-27. DOI: 10.1016/j.procbio.2016.12.020

[79] Bhadani A, Kafle A, Ogura T, Akamatsu M, Sakai K, Sakai H, et al. Current perspective of sustainable surfactants based on renewable building blocks. Current Opinion in Colloid & Interface Science. 2020;**45**:124-135. DOI: 10.1016/j.cocis.2020.01.002

[80] Ruksana J, Andrew MB, Marina T, Paschalis A. Biosurfactants, natural alternatives to synthetic surfactants: Physicochemical properties and applications. Advances in Colloid and Interface Science. 2020;**275**:102061. DOI: 10.1016/j.cis.2019.102061

Chapter 4

Modification of Clay Minerals by Surfactant Agents: Structure, Properties, and New Applications

Safa Gamoudi and Ezzeddine Srasra

Abstract

Organically surfactant-modified clays (OC) have attracted a great deal of interest because of their wide applications in industry and environmental protection. The OC are organic–inorganic complexes synthesized through the intercalation of organic cations mainly into the interlayer space of expandable clays. Different surfactants have been used to prepare OC. These include single and dual-cationic surfactants, anionic–cationic surfactants, and nonionic surfactants. The intercalation of the surfactant cations was governed by different processes: cationic exchange and Van der Waals 'interactions of the alkyl chains with clay surface. The structure and properties of the resultant organoclays are affected by the type of surfactant, the clay used, and the preparation method such as the conventional technique, the solid-state intercalation, and the microwave irradiation. As the result, the organoclays are characterized by hydrophobic surfaces and have attracted great interest because of their potential use in several applications, such as sorbents for organic pollutants (dyes, pharmaceutic compounds ...), heavy metals and inorganic oxy-anions, clay-based nanocomposites, and in several other industries.

Keywords: surfactant, modification, clay, structure, application

1. Introduction

Intercalation of organic guest species into layered inorganic solids is an interesting topic from a theoretical and applied point of view [1]. At present, there are many applications of organoclays used as sorbents in pollution prevention and environmental remediation such as the treatment of oil spills, wastewater and hazardous waste landfills, etc. Previous studies have shown that replacing the inorganic exchangeable cations of clay minerals with organic cations can greatly enhance the capacity of these materials to remove organic contaminants [2, 3]. Organoclay-based nanocomposites exhibit a remarkable improvement in surface properties when compared with untreated polymer or conventional micro- and macro-composites. These improvements include increased strength and heat resistance, decreased gas permeability and flammability, and increased biodegradability of biodegradable polymers [4, 5].

Different surfactants have been used to prepare organoclays. These include single and dual cationic surfactants [6–8], anionic–cationic surfactants [9] and nonionic surfactants [10]. The prepared organoclays, however, are structurally different even when the same surfactant was used under similar experimental conditions [11–14]. Hence, the structure and properties of the resultant organoclays are affected by the type of surfactant [15], the clay mineral used and the preparation method.

2. Surfactants

2.1 Definition

Surfactants are natural or synthetic substances whose molecules have a so-called amphiphilic structure (**Figure 1**). These molecules thus have i) a polar head, ionized or not, capable of developing Van der Waals, Lewis acid-base, and possibly Coulomb-type interactions when an ionizable function is present: this hydrophilic part has an affinity for charged surfaces and for liquids with a strong polar character such as water; and ii) an apolar part, typically a hydrocarbon chain capable only of Van der Waals interactions and therefore having little affinity for water.

2.2 Classification of surfactants

Surfactants are historically divided according to their charge into four main classes, cationic, anionic, amphoteric and non-ionic.

- Cationic surfactants: having a positively charged polar head, belong either to the family of fatty amines such as amines, diamines and polyamines as well as their salts (quaternary ammonium salts or imidazolinium salts) or to the family nitrogen heterocycles.

- Anionic surfactants: having a negatively charged polar head, are generally alkylarylsulphonates, alkane sulphonates, olefin sulphonates, salts of carboxylic acids (soaps), salts of sulfuric acid esters (alkyl sulphates), alkyl ether sulphates, and esters of phosphoric and polyphosphoric acids.

- Amphoteric surfactants: So-called amphoteric surfactants such as alkylamino acids, alkyl betaines and sulfobetaines, have two functional groups, one anionic and the other cationic. Depending on the environmental conditions, they can

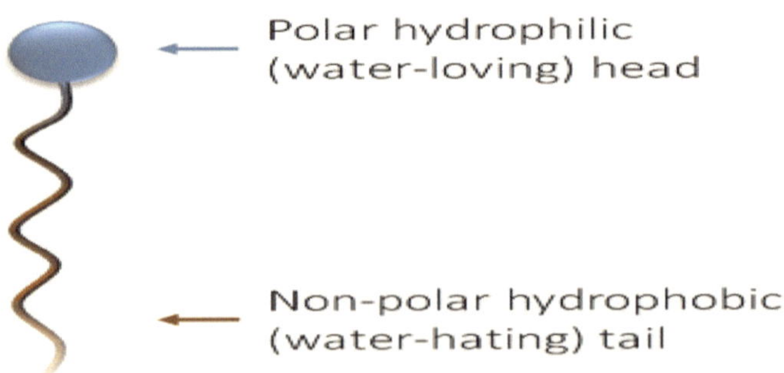

Figure 1.
Surfactant structure.

ionize in aqueous solution, acquiring an anionic character at alkaline pH and a cationic character at acidic pH.

- Nonionic surfactants: are compounds whose hydrophilic group is a chain of ethylene oxide units, generally attached to a hydroxyl function: alcohols, alkylphenols, carboxylic acids and alkanolamides.

2.3 Physicochemical proprieties

2.3.1 Critical micellar concentration (CMC)

The CMC is the concentration in the solution of a surfactant above which part of the molecules dispersed within the solution come together to form micelles. Micelles are formed when the hydrophobic portions, unable to form hydrogen bonds in the aqueous phase, create a large increase in the free energy of the system. One way to lower this energy is by isolating the hydrophobic part of the water by adsorption onto organic matrices or forming micelles. Indeed, in the micelles, the hydrophobic parts group together towards the center, and the hydrophilic parts remain in contact with water. Surfactant micelles arrange in different spherical, globular or cylindrical microstructures, but spherical and irregular vesicles, tubular bilayers or lamellar structures (**Figure 2**) are most often encountered.

The CMC of a surfactant varies with its structure, the temperature of the solution, the presence of electrolytes or organic compounds. The effects of electrolytes on CMC are more pronounced for ionic surfactants. The variation in the size of the

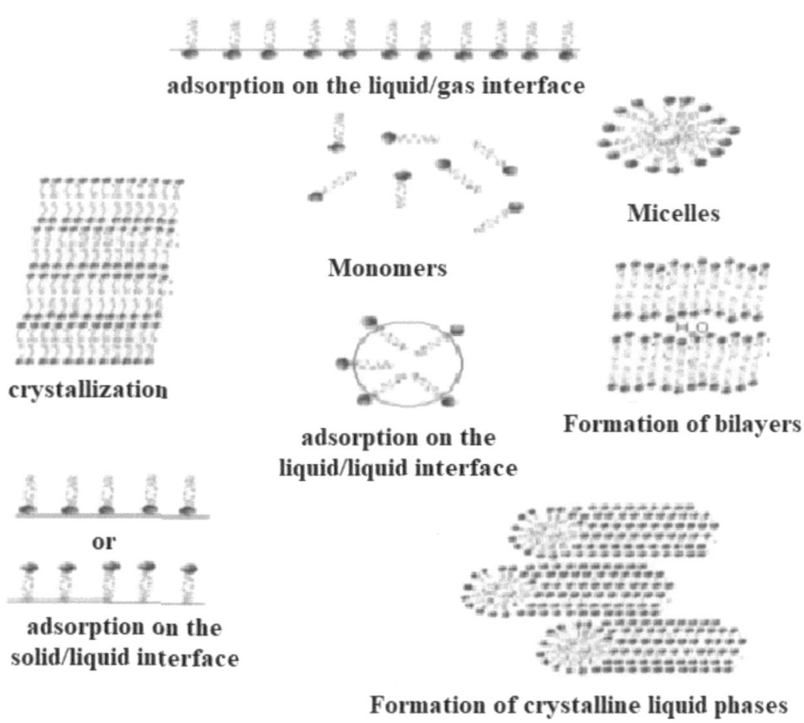

Figure 2.
Modes of aggregation and adsorption of surfactants [16].

hydrophobic region is an important factor and in general the CMC decreases as the hydrophobic character of the surfactant increases.

2.4 Adsorption at interfaces

Adsorption is a surface phenomenon that originates from the non-compensation in all directions of intermolecular attractions at interfaces. This results in residual inwardly directed forces, which are attenuated when amphiphilic entities attach to the surface. Information on this phenomenon is obtained by determining the surface tension of the solutions and the thermodynamic quantities of adsorption which vary according to the temperature.

3. Clays

3.1 Structures

Generally, clays are minerals composed mainly of hydrated aluminum or/and magnesium phyllosilicates. As their name suggests, they are formed by the association of layers. The latter consists of an assembly of layers of $Al(OH)_6$ "O" octahedra and layers of SiO_4 "T" tetrahedra. The tetrahedral layer T is formed by two planes of oxygen atoms and contains a silicon atom in IV coordination. In the tetrahedron, each silicon atom is surrounded by 4 oxygens. One tetrahedron is bonded to another through a highly covalent bond through the sharing of oxygen atoms. These are called basal or basal surface oxygens. The arrangement of these basal oxygens leads to the formation of hexagonal cavities. Opposite the basal surface are the apical oxygens which are shared between tetrahedral silicon and an octahedral cation. They establish a strong connection between the octahedra and the tetrahedra. The octahedral layer O is formed by two planes of oxygen atoms and hydroxyl groups between which aluminum or iron or magnesium cations are in VI coordination (**Figure 3**).

Phyllosillicates generally present isomorphic substitutions of Al^{+3} cations by Mg^{+2} and of Mg^{+2} by Li^{+} in the octahedral layers and/or Si^{+4} by Al^{+3} or Fe^{+3} in the tetrahedral layers. These substitutions lead to a charge deficit in the sheet, which becomes negatively charged. The electroneutrality of the structure is then ensured by the presence of compensating cations (Na^{+}, Ca^{+2}, K^{+}, etc.), which take place in the interfoliar space separating two layers. The thickness of the sheets depends on their composition. A distinction is made between clays of the 1:1 or TO type (where a tetrahedral T layer is linked to an octahedral O layer) and clays of the 2:1 or TOT type (in which an O layer is inserted between two T layers). For TO sheets, the thickness is 7 Å because the interfoliar space is empty. On the other hand, for TOT sheets it is variable from 9 to 15 Å depending on the content of the interfoliar layer (relative humidity, solvation liquid, nature and hydration of the interfoliar cation). In addition, clays may have other morphologies, such as nanotubule halloysite with a diameter of 50 nm, an inner lumen of 15 nm and a length of 600–900 nm (**Figure 4**).

3.2 Classification

There are different classifications of clays. The most classic is based on the thickness and structure of the sheet. The first classification, established by the

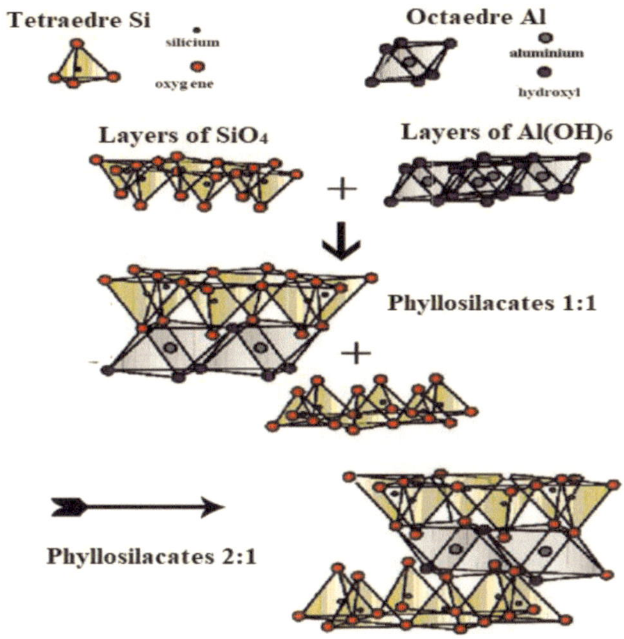

Figure 3.
Representation of silicon tetrahedra and aluminum octahedra and their layering and sheet arrangement [17].

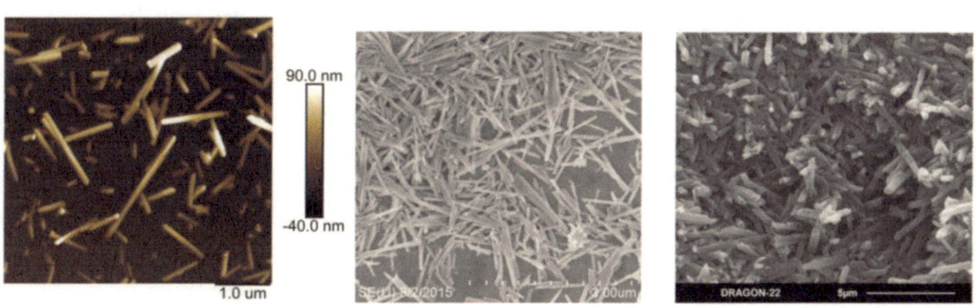

Figure 4.
Images of halloysite [18].

International Committee on Classification and Nomenclature of Clay Minerals in 1966, is based solely on the charge of the sheet and the number of metal atoms in the octahedral layer. The second, established by Mering and Pedro [19], takes into account the location of substitutions, their distribution and the type of compensating cations. This classification does not take into account synthetic silicates, sometimes used in the development of nanocomposites, such as fluorohectorite, fluoromica or laponite. **Table 1** presents a classification derived from the work of Brindley [20] and McKenzie [21] which gives the value of the permanent load of the sheet that was used as a criterion for establishing a classification of phyllosilicates 2:1.

In nature there are clay minerals formed by the succession of sheets, two or more types of clay, described above. These interstratified clays may have a simple or irregular regular interstratification where no law governs the alternation of sheets.

Type	Family	Charge of O_{10} (OH) $_2$ unity	octahedral Olayer	Exemples
TO	Kaolins	x ≈ 0	Dioctahedral	Kaolinite, dickite, nacrite
	Serpentines		Trioctahedral	Chrysotile, antigorite, berthiérine
TOT	Pyrophillites	x ≈ 0	Dioctahedral	Pyrophillite
	Talcs		Trioctahedral	Talc, willemséite
	Smectites	x from 0.2 to 0.6	Dioctahedral	Montmorillonite, beidellite, nontronite
			Trioctahedral	Saponite, stevensite
	Vermiculites	x from 0.6 to 0.9	Dioctahedral	
			Trioctahedral	
	Micas	x from 0.9 to 1.0	Dioctahedral	Muscovite, illite
			Trioctahedral	Phlogopite, biotite

Table 1.
Classification of phyllosilicates.

3.3 Proprieties of clays

3.3.1 Swelling capacity

The swelling consists of a separation of the sheets up to an interfoliar distance of equilibrium. This swelling property is due to the hydrophilic nature of the surface of certain clays, due to the presence of hydratable cations in the interfoliar spaces [22]. The intensity of the swelling depends on the charge of the crystal lattice, the nature of the compensating cations, the hydration energies involved, the ionic strength of the surrounding medium and the total quantity of water. The ability to swell is therefore not a characteristic common to all phyllosilicates. In the case of micas, the location of the isomorphic substitutions in the tetrahedral layer as well as the strong charge deficit create strong bonds between the compensating ions and the sheets; which prevent the hydration of the cations. In smectites, octahedral substitutions promote swelling because interactions between sheets and compensating cations are reduced. Smectites have the best swelling properties. This property is used as an identification criterion.

3.3.2 Cation exchange capacity CEC

The cation exchange capacity, denoted CEC, corresponds to the number of monovalent cations that can be substituted for the compensating cations of the clay. The latter are localized on the external surfaces of the particles as well as between the unit sheets [23]. The CEC is generally expressed in milliequivalents per 100 grams of calcined clay (meq/100 g). It varies according to the minerals, it is between 80 and 150 meq/100 g for smectites.

3.3.3 Specific surface area

In the anhydrous state, the clay sheets are joined to each other, but they move apart in the presence of water (swelling), which makes the basal surfaces initially in

Clay	Internal surface ($m^2.g^{-1}$)	External surface ($m^2.g^{-1}$)	SSA ($m^2.g^{-1}$)
kaolinite	0	10–30	10–30
illite	20–55	80–120	100–175
Smectite	600–700	80	700–800
chlorite	—	100–175	100–175

Table 2.
Specific surface area (SSA) of some clay minerals [24].

contact accessible. These constitute the inner surface of the mineral. The outer basal surfaces and the edge surfaces of the sheets constitute the outer surface of the clay. **Table 2** presenting the surface values of some clay minerals shows that smectites have the highest total surface. They have an internal surface of around 700 $m^2.g^{-1}$, while the external surface only represents around 80 $m^2.g^{-1}$.

3.3.4 Acid-base properties of clays

Surface charge is a fundamental property of clays. The main origin of this surface charge comes from isomorphic substitutions within the crystal lattice. This negative charge is commonly denoted permanent structural charge (σ_0) [25] and is around 7.10^{-3} e/Å^2. The second source of charge, which depends on the pH (denoted σ_H), comes from the presence of oxide or hydroxyl groups of the silanol (-SiOH) and aluminol (-AlOH) types. These hydroxyl groups, present at the edges of the surfaces of the clays, admit an amphoteric character according to the reactions:where $\equiv SOH^+{}_2$, $\equiv SOH^0$ et $\equiv SO^-$ represent the hydroxide groups of the edges' surfaces. K_1 and K_2 are the constants of protonation and group deprotonation reactions $\equiv SOH^0$.

The surface groups can therefore carry negative, positive or neutral charges and thus generate a charge on the surface of the particle. This charge is directly related to surface acid–base balances and/or more general surface adsorption reactions. It, therefore, depends on the composition of the liquid phase, mainly on the pH and the ionic strength. There is a particular pH value for which the proportions $\equiv SOH^+{}_2$ and $\equiv SO^-$ are equivalent: this is the point of zero charges (noted PCN). The point of zero charge PCN corresponds to the pH for which the global surface charge density σ is zero and therefore the activities $\equiv SOH^+{}_2$ and $\equiv SO^-$ are equal.

4. Clay modification by surfactant agents

The first published results on organophilic clays appeared in the early 1940s. Bradley [26] studied the adsorption of a series of polyamines and polyalcohols on certain types of montmorillonites. The advanced work of Jordan [27] showed that the intercalation of montmorillonites by quaternary ammoniums converses a hydrophilic character. In 1962, Fripiat et al. [28] used sodium, calcium and acid homoionic montmorillonites for the adsorption of certain amines (monoamines and diamines); they showed that acid montmorillonite adsorbs these products more than the other two matrices. Later, McBride et al. [29] as well as Karichoff et al. [30] showed that it was possible to use organophilic clays for the adsorption of certain aromatic compounds.

From the 1980s, many studies [31–40] on the interactions between several surfactants and different clays were carried out by examining several parameters such as the

Family	Surfactant	Structure	Reference
Ammonium quaternaries			
Cationic	Cethyltrimethylammonium (CTA)	$C_{19}H_{42}N^+$	Pospisil et al. [41] Juang et al. [42] Praus et al. [43]
	Hexadecyl-trimethyl-ammonium (HDPy)	$C_{22}H_{38}N^+$	Dultz and Bors [44] Janek and Lagaly [45] Ni et al. [46]
	Hexadecyl-trimethyl-ammonium (HDTMA)	$C_{19}H_{42}N^+$	Sheng and Boyd [47] Lee and kim [48] He et al. [12]
	Octadecyltrimethyl-ammonium (ODTMA)	$C_{21}H_{46}N^+$	Xi et al. [14] Hrachová et al. [49] Sanchez-Martin et al. [50]
	Benzyltrimethylammonium (BTMA)	$C_{10}H_{16}N^+$	Polubesova et al. [51] Shen [52] Park et al. [53]
	Tetrabutylammonium (TBAM)	$C_{16}H_{36}N^+$	Akcay et al. [54]
Anionic	Dodecylsulfate (SDS)	$C_{12}H_{25}O_4S^-$	Permien and Lagaly [55] Günister et al. [56]
Amphoteric	Triton X-102	$C_{16}H_{26}O_2$	Lin et al. [57]
	Cocamidopropyl-betaine (CAB)	$C_{19}H_{38}N_2O_3$	McLauchlin and Thomas [58]
	Polyethylene-glycol (PEG-1500)	$H(OCH_2CH_2)_nOH$	Zampori et al. [59]
Phosphonium quaternaries			
Cationic	Tetrabutylphosphonium (TBP)	$C_{16}H_{37}OP^+$	Hedley et al. [60]
	Tetraphenylphosphonium (TPP)	$(C_6H_5)_4P^+$	Patel et al. [61] Jeschke and Meleshyn [62]
	Tributyltetradecylphosphonium	$C_{26}H_{56}P^+$	Hedley et al. [60] Calderon et al. [63]
	Triphenyl vinylbenzyl phosphonium Tetraoctyl phosphonium	$C_{25}H_{22}P$ $C_{32}H_{68}P^+$	Avalos et al. [64]

Table 3.
Principal families of used surfactants.

TA/clay ratio, the pH of the environment and the nature of the intercalated surfactant. **Table 3** presents the main families of surfactants used for the modification of smectites.

4.1 Modification methods

Several elaboration strategies have been used with success to obtain organophilic clays, it is possible to distinguish the two most common modes of implementation: intercalation and grafting. In this part, we will describe in detail the first mode by presenting different experimental techniques of intercalation. This process consists of replacing the compensating cations with cations bearing alkyl chains. The most frequently used cations are alkylammonium ions. Phosphonium salts are also interesting modifying ions because their thermal stability is high but they are little used until now.

4.1.1 Batch method

This method of preparation is the most common, it consists of mixing the surfactant and clay in a common solvent and then removing it. In order to optimize this method it is necessary to use a solvent that can both swell the clay and dissolve the surfactant. The swelling of the clay facilitates the insertion of alkylammonium ions into the interfoliar spaces. After filtration of the suspension and drying of the clay, the presence of alkylammonium ions on the surface of the sheets, primary particles, and aggregates gives the clay an organophilic character. In addition, their intercalation between platelets leads to an increase in the interfoliar distance. The main disadvantage of this method is the high amount of solvent used which poses an environmental problem.

4.1.2 Solid-state reaction

This method of preparation is the most common, it consists of mixing the surfactant and the clay in a common solvent, then removing it. In order to optimize this method, it is necessary to use a solvent that can both swell the clay and dissolve the surfactant. The swelling of the clay facilitates the insertion of the alkylammonium ions within the interfoliar spaces. After filtration of the suspension and drying of the clay, the presence of alkylammonium ions on the surface of the sheets, the primary particles, and the aggregates give the clay an organophilic character. In addition, their intercalation between the platelets leads to an increase in the interfoliar distance. The main drawback of this method is the large amount of solvent used which poses an environmental problem.

4.1.3 Micro-wave irradiation

The use of microwaves in the synthesis of organophilic clays presents a new preparation technique. It offers more advantages than conventional and thermal methods listing high preparation time and energy consumption. However, there are few published works on the use of microwave irradiations to prepare organophilic clays. In 2006, Baldassari et al. synthesized organophilic clays using the conventional method (liquid state intercalation) and synthesis by microwave activation in order to compare the results obtained by the two synthesis methods under the same conditions (amounts of surfactants, mass of clay, volume of solution). The authors have shown that the intercalation of surfactants is complete after 2 hours of irradiation. However, with the classical method, the intercalation was completed after a 6-hour treatment. These results prove that microwave treatment is more effective than conventional treatment. The same conclusions are found by Li et al. [65]. They showed that the synthesis of organophilic smectites under the effect of microwave irradiation is beneficial since 5 minutes of irradiation is sufficient for the intercalation of an amount equal to 70%CEC. The advantages of synthesis under the effect of microwave irradiation are also visualized by Korichi et al. [66], Gunawan et al. [67] and Liu et al. [68]. These authors concluded that the structural and surface properties are improved.

4.2 Surfactant intercalation mechanism in clays

From an experimental point of view, it is often difficult to highlight the nature of the bonds that can be established between solutes and a system as complex as clay, especially since, for the same adsorbent-surfactant couple, several connections of different natures can be established. However, it is essential to understand the adsorption mechanisms in order to correctly interpret the partition of a solute between the aqueous liquid phase and the solid phase of the clay. The techniques used to identify adsorption mechanisms are XRD, IR, SEM, BET analysis techniques [43] or thermodynamic techniques for measuring heats of adsorption by microcalorimetry [69]. In most works on the study of adsorption mechanisms, adsorption is described based on empirical models making it possible to simulate the global phenomena observed [70]. Despite all the difficulties encountered in describing the mechanisms of adsorption of organic pollutants in soils, two main categories of mechanisms can be distinguished.

4.2.1 Ion exchange

It has an important role in the adsorption of surfactants in clay, particularly in the case of ionizable molecules. It is formed between cations or anions and, respectively, negative or positive charges located on the surface of the adsorbent. The intensity of adsorption will closely depend on the solution pH, temperature and ionic strength.

4.2.2 London: Van der Waals connections

The attractive forces involved in this type of bonds correspond to dipolar interactions between the solvent (water), the solutes, and the solid surfaces. These are electrostatic forces due to the movement of electrons which change atomic orbitals. These are low binding energy interactions. Their intensity increases with the size of the molecules due to their additive nature. These are the main forces responsible for the physical adsorption of surfactants in clays.

4.3 Factors influencing intercalation

4.3.1 Influence of the clay proprieties

The quantity of modifying ions adsorbed on the surface of the layers depends on the number of accessible sites, therefore on the CEC and on the structure of the clay mineral. The isomorphic substitutions of montmorillonite are in the octahedral layers. Electrostatic interactions with compensating cations are therefore attenuated by the tetrahedral layer. It presents the most interesting compromise between a CEC large enough to allow a quality organophilic modification without sterically encumbering the interfoliar space, and small enough to allow the separation of the sheets in aqueous media.

4.3.2 Influence of exchanged cation

The inorganic compensating cation of the clay to be replaced also has an influence. This is related to its role during the dispersion of clay in aqueous solution. Larger and highly charged cations limit the opening of interfoliar spaces and are less easily exchangeable. The effect of the charge carried by the cation was also observed by McAtee [71]. He proved that for the same alkyl ammonium ion, a total exchange takes place with sodium montmorillonite while it remains limited to 70 or 80% with montmorillonites containing calcium or magnesium ions. In summary, the smaller and more mobile the compensating cation, the more the cation exchange is facilitated. The compensating cations most frequently present in clays can therefore be classified in increasing order of aid to cation exchange: $Cs^+ < Rb^+ < Ca^{2+} < Mg^{2+} < Na^+ < Li^+$.

4.3.3 Influence of the alkylammonium ion nature

The type of alkylammonium ion plays a considerable role in the intercalation. Indeed, the length of the carbon chain, the size and shape of the polar head, as well as the organic groups carried by the ion have significant influences on the efficiency of the exchange. The increase in the interlayer space, and therefore the improvement of the dispersion in a polymer and the properties of the final material, are related to the length of the carbon chain of the alkylammonium ion. Thus, by increasing the length of the carbon chain, we increase the entropic contribution of the adsorption energy and we develop more Van der Waals interactions. The binding of organic cations also depends on the size and shape of the polar head, according to Rowland and Weiss [72]. The results of work by McAtee [71] showed that ions from primary amines were not adsorbed in sufficient quantity to reach the CEC, unlike quaternary ammoniums. It has also been shown that the bond strength of amino derivatives decreases sharply from primary to secondary and tertiary compounds. Therefore, the groups carried by the carbon chain of the ion also affect the quality of the ion exchange. Indeed, cation exchange is favored when the ammonium cation presents a group capable of interacting with the surface oxygens of the sheets through hydrogen bonds.

4.4 Organoclay structures

It is not easy to control the structure obtained after the organophilic modification, because the chains can adopt different conformations within the interfoliar space. The type of arrangement obtained in these spaces is strongly dependent on the initial concentration of alkylammonium with respect to the CEC of montmorillonite. Indeed, the adsorption of the first layer of ions on the surface is linked to the cation exchange process, but the layers adsorbed thereafter are linked to the first by chain/chain interactions of the Van der Waals type and follow classical adsorption laws.

Lagaly, [34, 73] described the probable conformations of alkylammonium ions on the surface of the sheets. Depending on the length of the carbon chain and the charge deficit of the sheet, they can be organized in monolayers, in bilayers, according to a pseudotrimolecular or paraffinic type arrangement (**Figure 5**). Furthermore, Gherardi [74] described the organization of alkylammonium ions (of carbon chains with more than twelve methylene groups) in the context of adsorption above the CEC and its consequences on the multi-scale organization of montmorillonites in aqueous solution. The author observes that if the initial ion concentration is equal to the CEC of montmorillonite, the clay is completely hydrophobic and flocculates due to interactions between carbon chains. The interpretations and proposals for chain conformations made by Lagaly and Weis [73] use only X-ray diffraction measurements and are based on the assumption of the ideal case of an arrangement of the carbons of the alkyl chain in trans conformation. Vaia et al. [75] dispute these results based on infrared spectrometry measurements.

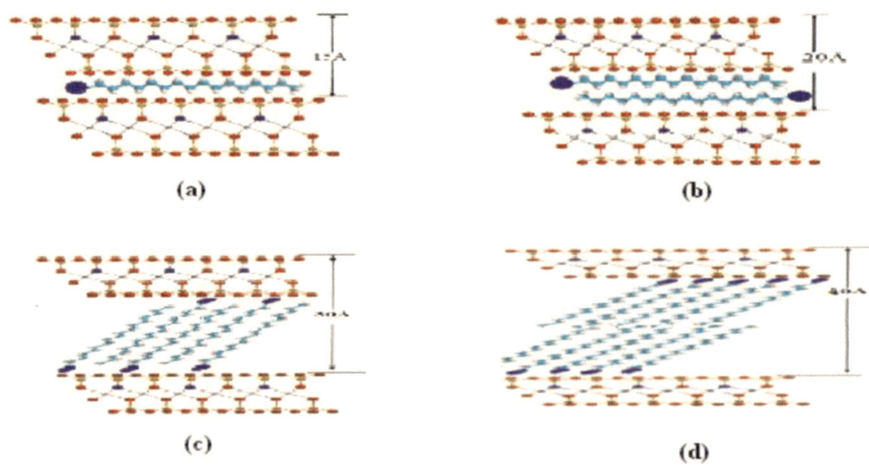

Figure 5.
Idealized structure of surfactants intercalated between the sheets of a phyllosilicate according to Lagaly [34]: (a) monolayer, (b) bilayer, (c) pseudotrimolecular, (d) paraffinic.

Contrary to the previous case which only considers trans conformations to establish their structures, here the values of the wavelengths as well as their variations associated with the vibrations of the methyl groups which can be in left or trans conformations are studied. This work shows that the intercalated chains can be in gas, liquid, liquid crystal or solid states depending on the length of the alkyl chains (**Figure 6**). They show in particular that the same interfoliar distance can come from different conformations. Hackett et al. [76] confirmed these interpretations by molecular dynamics simulations. It can be noted that work based on NMR experiments has also been carried out to determine the conformation of molecules and has shown the coexistence of trans-ordered and left-disordered conformations. Moreover, these authors suggested that the charge density of the clay and the length of the aliphatic chain of the intercalated surfactant are the two interesting parameters influencing the structure of the complex obtained.

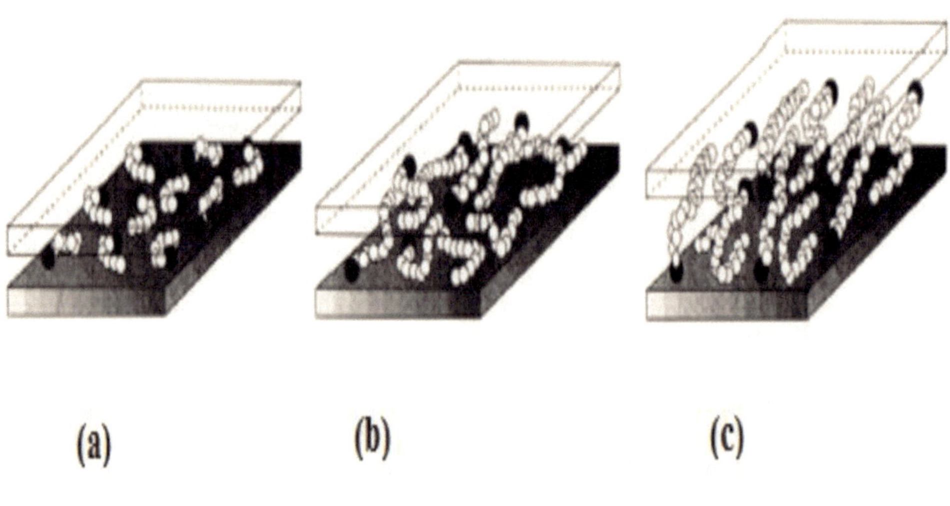

Figure 6.
Model conformations of chains of different lengths and intercalated according to Vaia [75]. (a) Short chains in C_6 (b) medium in C_{12} (c) long chains in C_{18}.

5. Applications of surfactant modified-clays

All the results obtained so far concerning the different interactions of organophilic clays with organic compounds such as benzene and toluene [77] and phenol and its derivatives [78] have shown that these new materials are effective adsorbents.

Moreover, certain other recent works on bridged clay-pesticide interactions (in particular herbicides) have confirmed the hydrophobic nature of these complexes. This is how Nahhal et al. [79] and Carrizosa et al. [80] have shown through the various results obtained that there are indeed strong pesticide-organophilic clay interactions.

In the field of discoloration of textile industry effluents, Elguendi [81], Choi and Cho, [82] were interested in the adsorption of certain dyes such as basic blue 69 (Artrozone blue) and basic red 22 (Maxillon red; $C_{24}H_{18}N_4O_4$) on montmorillonites and vermiculites inserted by surfactants. Overall, they observed strong adsorbent-adsorbate affinities attributed to the hydrophobic nature of the surfactant molecules inserted into these clays.

The organo-clays were attractive for use as selective sorbents due to the organic layer such as dyes like Acid Blue 193 [83], acid dyes [84], and basic dyes [85], Congo red [86], Acid Orange 7 and Acid Blue 9, Malachite green [87], and Basic Red 18 [88]. According to the results obtained by these authors, the organo-clay was more effective than raw clay in the removal of reactive, basic, or anionic dyes. This result is due to the fact that after modification, the surface area and its porosity also increase.

Recent studies by Oyanedel-Craver and Smith [89], Oyanedel-Craver et al. [90], Guimaraes et al. [91] on the treatment of water polluted by heavy metals using montmorillonites intercalated with cationic surfactants confirmed the organophilic nature of these matrices and made it possible to classify their adsorption capacities.

Jiang et al. [92] modified an MMT montmorillonite with solutions of Fe, Al, Fe/Al, Fe-HDTMA, Al-HDTMA and Fe/Al-HDTMA hydroxides. The X-ray analyzes of different modified clays showed an increase in the basal distance d_{001} from 15 to 18 Å, of which the Fe/Al-HDTMA-MMT clay presents the highest distance. The authors then used modified clays for the adsorption of phenol. The adsorption isotherms obtained show that the MMT treated with HDTMA is more effective for the retention of phenol molecules than the natural MMT and treated with Fe, Al, Fe/Al.

Oyanedel-Craver et al. [90] studied the simultaneous retention of benzene and heavy metals (Pb, Cd, Zn and Hg) by MMTs modified with HDTMA and BTEA. They showed that metal retention is achieved by the formation of complexes with silanol and aluminol groups. By comparing the results of benzene adsorption by HDTMA-MMT and BTEA-MMT, the authors concluded that the mechanism of retention differs according to the length of the carbon chain of the intercalated surfactant: if the chain is short, the retention takes place by interaction between the surface of the clay and the organic molecule whereas if the chain is long, the retention is done by interaction between the surfactant chain and the organic molecule.

Szanto et al. [93] examined the adsorption of methanol and benzene by two montmorillonites sodium and calcium modified by different amounts of surfactant HDPy (20–90 mmol/100 g). They showed that the methanol and benzene retention capacity is higher for organo-clays than for natural clays. The fixed amounts of methanol and benzene on natural and modified sodium clay are 3.44; 1.8; 8.25 and 4.35 $mmol.g^{-1}$.

Tvardovski et al. [94] studied the retention of benzene and hexane vapors by a natural Montmorillonite clay (CEC = 100 meq/100 g, SAA = 34 m^2/g) and a synthetic fluorohectorite clay (CEC = 82 meq/100 g, SAA = 12 m^2/g) modified by HDPy. The authors showed that the retention of benzene (Q_{ads} = 5.2 $mmol.g^{-1}$ by HDPy-MMt, Q_{ads} = 2.6 $mmol.g^{-1}$ by HDPy-fluorohect) is greater than that obtained for hexane (Q_{ads} = 0.72 $mmol.g^{-1}$ by HDPy-MMt, Q_{ads} = 0.78 $mmol.g^{-1}$ by HDPy-fluorohect). By comparing the adsorption results of benzene and hexane, the authors concluded that the retention mechanisms differ according to the microstructure and the crystallite of the organo-clays prepared.

Bors et al. [95] used a sodium bentonite (CEC = 76 meq/100 g) modified by quantities of HDPy varying from 0.0 to 2.0 CEC as an adsorbent of iodine (I), technetium (Tc), cesium (Cs) and strontium radionuclides (Sr). They showed that the quantities of radionuclides adsorbed by organophilic bentonite exceed those adsorbed by natural bentonite and also that the iodide ions are more adsorbed (I˃Te˃Ce˃Sr). This adsorption is influenced by the structure of the modified clay and the intercalated quantity of the surfactant. They found that these radionuclides are retained primarily by ion exchange.

Behnsen and Reibe [96] modified a montmorillonite of CEC = 89 meq/100 g by three cationic surfactants HDTMA, BE and HDPy. The adsorption isotherms of I^-, NO_3^-, ReO_4^-, SeO_3^{-2} and SO_4^{-2} anions with an initial concentration of 1–10 $mmol.L^{-1}$ show good efficiency of these organo-clays in the retention of anions. The authors concluded that the anions are retained by anion exchange. The anion selective order is $I^- > ReO_4^- > NO_3^- > Cl^- > SO_4^{-2} > SeO_3^{-2}$ for the three organo-clays. Comparatively, the fixed amounts of I^-, ReO_4^-, NO_3^-, Cl^-, SO_4^{-2} and SeO_3^{-2} on the HDPy-MMt clay are the highest, they are respectively 452, 431, 393, 276, 88 and 75 $mmol.Kg^1$.

Yab et al. [97] were prepared inorganic tubular micelle. Selective modification of aluminosilicate clay nanotube inner lumen with octadecyl phosphonic acid and dopamine was demonstrated. The adsorption study showed that halloysite with organosilane adsorbs more ferrocene than its hydrophilic derivative (ferrocene carboxylic acid). Therefore, like in organic micelle, the octadecyl phosphonic acid immobilized in the halloysite lumen may behave as sponge for physisorption increasing adsorption capacity for hydrophobic molecules.

Owoseni et al. [98] proposed a new technology for the treatment of oil spills using organo-modified clay. The method is based on principles of emulsification using naturally occurring halloysite clay with structure $Al_2[Si_2O_5(OH)_4].2H_2O$ and the Dioctyl sulfosuccinate surfactants from the halloysite nanotubes to the oil-water interface. At appropriate surfactant compositions and loadings in halloysite nanotubes, the crude oil-saline water interfacial tension is effectively lowered to levels appropriate for the dispersion of oil (**Table 4**).

Retention of organic compounds			
adsorbate	**Surfactant modified clay**	**Results**	**Reference**
Phenol and its derivatives			
Phenol	MMT modified by Fe/Al-HDTMA	Q_{phenol} = 6 $mg.g^{-1}$	Jiang et al. [92]
	Bentonite modified by HDTMA	* Q_{phenol} = 180 $mol.g^{-1}$	Yapar et al. [99]
	Bentonite modified by HDTMA	oxidation 100	Mojovic et al. [100]

Retention of organic compounds			
adsorbate	**Surfactant modified clay**	**Results**	**Reference**
Pentachlorophenol PCP	Bentonite modified by (DODMA, HDTMA, HDPy, TMA)	Q_{PCP}(bent-$_{HDTMA, DODMA}$) = 0.12 mg.g^{-1} Q_{PCP}(bent-$_{TMPA,TMA}$) = 0.02 mg.g^{-1}	Boyd et al. [101]
	Na-MTM modified by HDTMA	$Q_{PCP\ max}$ = 0.6 mol/kg	Stapleton et al. [102]
A = 3, 4, 5-trichlorophénol, B = 3, 5-dichlorophénol C = 3-chlorophénol	Bentonite modified by HDTMA	Q_{adsA} = 2 mmole/g Q_{adsB} = 3 mmole/g Q_{adsC} = 0.9mmole/g	Mortland et al. [103]
Tanin	Bentonite modified by HDTMA	$Q_{Tanin\ max}$ = 2.5	
para-nitrophénol PNP	Ca-MMT modified by HDTMA	—	Zhou et al. [78]
2,4,5-trichlorophénol	Bentonite modified by HDTMA	Q_{adsA} = 374.9 mg/g	Zaghouane-Boudiaf and Boutahala [104]
Benzene and its derivatives			
Benzene BZ	Smectites modified by HDPy	Q_{BZ} = 5.5 mmol.g^{-1}	Stul et al. [38]
Methanol M Benzene BZ	Na-MMT modified by HDPy	Q_M = 8.25 mmol.g^{-1} Q_B = 4.35 mmol.g^{-1}	Szanto et al. [93]
Nitrobenzène NB Trichloréthylène TCE	Smectite modified by HDTMA	Q_{NB} = 2 mmol.g^{-1} Q_{TCE} = 4 mmol.g^{-1}	
Benzene BZ Trichloro-ethane TCE 1,2-dichloro-benzene DCB	Bentonite modified by HDTMA	Q_{DCB} = 700 mg.kg^{-1} Q_{BZ} = 600 mg.kg^{-1} Q_{TCE} = 400 mg.kg^{-1}	Bartelt-Hunt et al. [105]
Nitrobenzène	Na-MMT modified by HDTMA	Q_{NB} > 100 mmol.kg^{-1}	Borisover et al. [106]
Pesticides			
Triadimefon TDF	MMT (SW, SA) modified by HDTMA	Q_{TDF} = 6	Celis et al. [107]
Sulfometuron SFM	Na-MMT modified by HDTMA	Taux(%) = 60.0%	Mishael et al. [108]
Fluridone FD	Na-MMT modified by HDTMA	Q_{FD}(MMT $_{HD-0,8CEC}$) = 35 μmol.g^{-1}	Yaron-Marcovich et al. [109]
allelochemical scopoletin	commercial organoclay	—	Galán-Pérez et al. [110]
Pharmaceutical wastes			
lactam antibiotics	MMT modified by DDAB	Removal 97%	Saitoha and Shibayama [111]

Retention of organic compounds			
adsorbate	**Surfactant modified clay**	**Results**	**Reference**
diclofenac sodium	Bentonite modified DMA	Qmax = 0.115(mmol/g)	Maia et al. [112]
codeine, oxazepam, ibuprofen	MMT modified by BDTA and Brij-O20	R (BDTA-MMT) = 12.74% R (Brij-O20-MMT) = 59.38% R (BDTA-MMT) = 81.99% R (Brij-O20-MMT) = 48.98% R (BDTA-MMT) = 92.74% R (Brij-O20-MMT) = 47.64%	De Oliveira et al. [113]
AZM antibiotic	L-methionine modified montmorillonite	Qmax = 298.78 mg/g	Imanipoor et al. [114]
Dyes and pigments			
IC dye	PDADMA-bentonite	Qmax = 149.2 mg/g	Shen et al. [115]
	$HDPy^+$-clay	Qmax = 326.31 mg/g	Gamoudi and Srasra [116]
MO dye	HDTMA-bentonite	Qmax = 141.0 mg/g	Belhouchat et al. [117]
	$HDPy^+$-clay	Qmax = 277.27 mg/g	Gamoudi and Srasra [116]
RP dye	CP-montmorillonite	Qmax = 19.78 mg/g	Lee et al. [118]
	$HDPy^+$-clay	Qmax = 344.82 mg/g	Gamoudi and Srasra [116]
CV dye	HDTMA-bentonite	Qmax = 162.5 mg/g	Anirudhan and Ramachandran [85]
	$HDPy^+$-clay	Qmax = 65.78 mg/g	Gamoudi and Srasra [119]
b-carotene	Bentonite modified by RSB	Qmax = 97.58 mg/g	Santoso et al. [120]
Retention of inorganic compounds			
Retention of anions			
iodure (I), technetium (Tc), cesium (Cs), strontium (Sr)	Na-Bent modified by HDTMA	Selectivity: I˃Te˃Ce˃Sr	Bors et al. [95, 121]
iodure I	Na-bent modified by HDPy	$K_d = 297\ L.Kg^{-1}$	Dultz et al. [122]
chromate	Am-Bent by HDTMA	$^*Q_{max} = 795\ mmol.Kg^{-1}$	Krishna et al. [123]
	MMT by HDTMA HDP	$^*\ Q_{ads} = 18\ mg.g^{-1}$	Brum et al. [124]
I^-, ReO_4^-, NO_3^-, Br^-, SO_4^{-2}, SeO_3^{-2}	Bent modified by HDTMA	$Q_{ads}\ (ReO_4^-) = Q_{ads}\ (I^-) = 1{,}5Q_{ads}(NO_3^-) = 2\ Q_{ads}\ (Br^-) = 8Q_{ads}\ (SO_4^{-2}) = 8Q_{ads}\ (SeO_3^{-2}) = 450\ mmol.kg^{-1}$	Behnsen and Riebe [96]

Retention of organic compounds			
adsorbate	**Surfactant modified clay**	**Results**	**Reference**
		Selectivity: ReO_4^{-}>I^{-} >NO_3^{-}>Br^{-}>Cl^{-}> SO_4^{-2} > SeO_3^{-2}	
Nitrate	Bent modified by HDTMA	Q_{ads} = 14,76 mg.g^{-1}	Xi et al. [125]
NO3-	Smectite/illite modified by HDTMA	Q_{NO3-} = 15.38 mg.g^{-1}	Gamoudi et al. [126]
F-	Smectite modified by HDPy	Q_{F-} = 19.34 mg.g^{-1}	Gamoudi et al. [127]
I^{-}, IO_3^{-}	Organoclay	Q_{ads} (I^{-}) = 12.1 mg.g^{-1}	Li et al. [128]
Adsorption of metallic ions			
Pb, Cd, Zn et Hg	Ca-bent modified by BTEA	Q_{Cd} = 7.82 mg.g^{-1} Q_{Pb} = 87,8 mg.g^{-1}	Oyanedel-Craver, [89, 90]
Sb (III)	Montmorillonite modified by HDPy	Q_{Sb} = 108.7 mg.g^{-1}	Bagherifam et al. [129]
Retention of gases			
CO, CH_4, SO_2	Bent modified by BTEA	Q_{SO2} = 1.65 mmol.g^{-1}	Volzone et al. [130]
O_2, N_2, CO, CH_4, C_2H_2, CO_2 and SO_2	Na^{+}-MMT modified by HDTMA	Q_{SO2} = 1.7 mmol.g^{-1}	Volzone [131]
NH_3 and CO_2	Laponite modified by PAMAM	Selectivity CO_2 > NH_3	Kinjal et al. [132]
CO_2	Halloysite d by HKUST-1	—	Park et al. [133]
Antimicrobial activity			
Salmonella enteritidis	SAz and SWy MMTs modified byHDPy	Q_{ads}(SAz-1-HDPy) = 97% Q_{ads}(SWy-2HDPy) = 95%	Herrera et al. [134, 135]
Escherichia coli	Montmorillonite modified by HDPy	Q_{ads} (SAz-1-HDPy) = 95%	Ma et al. [136]
Staphylococcus aureus *Escherichia coli*	Montmorillonite modified by TBA, DDA and TBP		Bujdáková et al. [137]
Bacillus subtilis	Hallyosite modified by CTAB and SDS		Abhinayaa and Mangalaraj [138]

Table 4.
Applications of surfactant modified clay.

6. Conclusion

The OC are organic–inorganic complexes synthesized through the intercalation of organic cations mainly into the interlayer space of expandable clays. Different surfactants have been used to prepare OC. These include single and dual cationic surfactants, anionic–cationic surfactants and nonionic surfactants. The intercalation of the surfactants cations was governed by different processes: cationic exchange and Van der Waals 'interactions of the alkyl chains with clay surface. The structure and properties of the resultant organoclays are affected by the type of surfactant, the clay used and the preparation method such as the conventional technique, the solid-state intercalation, and microwave irradiation. As a result, the organoclays are characterized by hydrophobic surfaces and have attracted great interest because of their potential use in several applications, such as sorbents for organic pollutants, heavy metals and inorganic oxy-anions, in clay-based nanocomposites, and in several other industries.

Author details

Safa Gamoudi[1,2*] and Ezzeddine Srasra[1]

1 Laboratory of Composite Materials and Clay Minerals, National Center for Research in Materials Science, Soliman, Tunisia

2 National Engineering School of Gafsa, University of Gafsa, Tunisia

*Address all correspondence to: safagamoudi@gmail.com

References

[1] Khaorapapong N, Ogawa M. In situ formation of bis(8-hydroxyquinoline) zinc(II) complex in the interlayer spaces of smectites by solid–solid reactions. Journal of Physics and Chemistry of Solids. 2008;**69**:941-948

[2] El-Nahhal Z, Safi M. Adsorption of phenanthrene on organoclays from distilled and saline water. Journal of Colloid and Interface Science. 2004;**269**: 265-273

[3] Han KC, Fan C, Chiang P, Wang M, Lin K. p-Nitrophenol, phenol and aniline sorption by organo-clays. Journal of Hazardous Materials. 2007;**149**: 275-282

[4] Liu X, Wu Q. PP/clay nanocomposites prepared by grafting–melt intercalation. Polymer. 2001;**42**: 10013-10019

[5] Tang Y, Hu Y, Song L, Gui Z, Chen Z, Fan W. Preparation and thermal stability of polypropylene/montmorillonite nanocomposites. Polymer Degradation and Stability. 2003;**82**:127-131

[6] IJdo WL, Pinnavaia TJ. Staging of organic and inorganic gallery cations in layered silicate heterostructures. Journal of Solid-State Chemistry. 1998;**139**: 281-289

[7] Wang CC, Juang LC, Lee CK, Hsu TC, Lee JF, Chao HP. Effects of exchanged surfactant cations on the pore structure and adsorption characteristics of montmorillonite. Journal of Colloid and Interface Science. 2004;**280**:27-35

[8] Yilmaz N, Yapar S. Adsorption properties of tetradecyl- and hexadecyltrimethylammonium bentonites. Applied Clay Science. 2004; **27**:223-228

[9] Zhu LZ, Chen BL. Sorption behavior of p-nitrophenol on the interface between anion–cation organobentonite and water. Environmental Science and Technology. 2000;**34**:2997-3002

[10] Shen YH. Preparations of organo-bentonite using nonionic surfactants. Chemosphere. 2001;**44**:989-995

[11] Baldassari S, Komarneni S, Mariani E, Villa C. Microwave versus conventional preparation of organoclays from natural and synthetic clays. Applied Clay Science. 2006;**31**:134-141

[12] He H, Frost R, Bostrom T, Yuan P, Duong L, Yang D, et al. Changes in the morphology of organoclays with HDTMA surfactant loading. Applied Clay Science. 2006;**31**:262-271

[13] Lee S, Kim S. Adsorption of naphthalene by HDTMA modified kaolinite and halloysite. Applied Clay Science. 2002;**22**:55-63

[14] Xi Y, Martens W, He H, Frost RL. Thermogravimetric analysis of organoclays intercalates with the surfactant octadecyltrimethylammonium bromide. Journal Thermal Analysis and Calorimetry. 2005;**81**:91-97

[15] Heinz H, Vaia RA, Krishnamoorti R, Farmer BL. Self-Assembly of Alkylammonium Chains on Montmorillonite: Effect of Chain Length, Head Group Structure, and Cation Exchange Capacity. Chemistry of Materials. 2007;**19**:59-68

[16] Manet S. Effet de contre-ion sur les propriétés d'amphiphiles cationique, thèse. France: l'université de bordeaux; 2007. p. 7

[17] Lantenois S. Réactivité fer métal/ smectites en milieu hydraté à 80°C, thèse

de doctorat, l'université d'Orleans. Ecole doctorale sciences et technologie. 2003. p. 13

[18] Lvov Y, Wang W, Zhang L, Fakhrullin R. Halloysite clay nanotubes for loading and sustained release of functional compounds. Advanced Materials. 2016;**28**:1227-1250

[19] Mering J, Pedro G. Discussion à propos des critères de classification des phyllosilicates 2/1. Bulletin du groupe français des argiles. 1969;**21**:1-30

[20] Brindley GW. Discussion and recommendations concerning the nomenclature of clay minerals and related phyllosilicates. Clays and Clay Minerals. 1966;**14**:27-34

[21] McKenzie RC. The classification of soil silicates and oxydes. In: Gieseking JE, editor. Soil Components (Inorganic Components). New York, Berlin, Heidelberg: Spring-Verlag; 1975. pp. 1-25

[22] Karpinski B, Szkodo M. Clay minerals – Mineralogy and phenomenon of clay swelling in oli and gas industry. Advanced in Material Science. 2015;**15**: 37-55

[23] Verbung K, Baveye P. Hysterisis in the binary exchange of cations on 2:1 clay minerals: A critical review. Clays and Clay Minerals. 1994;**42**:207-220

[24] Morel R. Les Sols Cultivés. Paris: Lavoisier; 1996

[25] Sposito G. Surface reactions in natural aqueous colloidal systems. Chimia. 1989;**43**:169-176

[26] Bradley WF. Molecular associations between montmorillonite and some polyfunctional organic liquids. Journal of American Chemical Society. 1945;**67**: 975-981

[27] Jordan JW. Organophilic bentonites I. swelling in organic liquids. Journal of Physic and Colloid Chemitry. 1949;**53**: 294-306

[28] Fripiat JJ, Servais A, Leonard A. Etude de l'adsorption des amines par la montmorillonite. Bulletin de la Société Chimique France. 1962:617-644

[29] McBride MB, Pinnavaia JJ, Mortland MM. Adsorption of aromatic molecules by clays in aqueous suspension. Advances in Environmental Science and Technology. 1977;**8**:145-154

[30] Karichoff SW, Brown DS, Scott TA. Sorption of hydrophobic pollutants on natural sediments. Water Resources Research. 1979;**13**:241-248

[31] Arroyo M, López-Manchado MA, Herrero B. Organo-montmorillonite as substitute of carbon black in natural rubber compounds. Polymer. 2003;**44**: 2447-2453

[32] Cetin K, Huf WD. Layer charge of the expandable component of illite/ smectite in K-bentonite as determined by alkylammonium ion exchange. Clays and Clay Minerals. 1995;**43**:150-158

[33] Kozak M, Domba L. Adsorption of the quaternary ammonium salts on montmorillonite. Journal of Physical and Chemical Solids. 2004;**65**:441-445

[34] Lagaly G. Interaction of alkyamines with differents types of layered compounds. Solid State Ionics. 1986;**22**: 43-51

[35] Majdan M, Maryuk O, Pikus S, Olszewska E, Kwiatkowski R, Skrzypek H. Equilibrium, FTIR, scanning electron microscopy and small wide angle X-ray scattering studies of chromates adsorption on modified bentonite.

Journal of Molecular Structure. 2005;**70**: 203-211

[36] Micheal AA, Frans RT, Cheok NT. Properties of water in calcium and hexadecyltrimethyl-exchanged bentonite. Clays and Clay Minerals. 1999;**47**:28-35

[37] Moraru VNM. Structure formation of alkylammonium montmorillonites in organic media. Applied Clay Science. 2001;**19**:11-26

[38] Stul MS, Van Leemput L, Leplat L, Uytterhoeven JB. The adsorption of organic Vapors on Alkylammonium Smectites. The influence of mineral charge density and Monofunctional ammonium cation type. Journal of Colloid and Interface Science. 1983;**94**: 154-165

[39] Xi Y, Ding Z, He H, Frost RL. Structure of organoclays—An x-ray diffraction and thermogravimetric analysis study. Journal of Colloid and Interface Science. 2004;**277**:116-120

[40] Zhang Z, Sparks D, Scrivner C. Sorption and desorption of quaternary amine cations on clays. Environmental Science Technology. 1993;**27**:1625-1631

[41] Pospisil M, Capkova P, Merınskay D, Malacy Z, Simonıky J. Structure analysis of montmorillonite intercalated with Cetylpyridinium and Cetyltrimethylammonium: Molecular simulations and XRD analysis. Journal of Colloid and Interface Science. 2001;**236**: 127-131

[42] Juang R, Lin S, Tsao K. Mechanism of sorption of phenols from aqueous solutions onto surfactant-modified montmorillonite. Journal of Colloid Interface and Science. 2002;**254**:234-241

[43] Praus P, Turicová M, Študentová S, Ritz M. Study of cetyltrimethylammonium and cetylpyridinium adsorption on montmorillonite. Journal of Colloid and Interface Science. 2006;**304**:29-36

[44] Dultz S, Bors J. Organophilic bentonites as adsorbents for radionuclides II. Chemical and mineralogical properties of HDPy-montmorillonite. Applied Clay Science. 2000;**16**:15-29

[45] Janek M, Lagaly G. Interaction of a cationic surfactant with bentonite: A colloid chemistry study. Colloid Polymer Science. 2003;**281**:293-301

[46] Ni R, Huang Y, Yao C. Thermogravimetric analysis of organoclays intercalated with thegemini surfactants. Journal of Thermal Analysis and Calorimetry. 2009;**96**:943-947

[47] Sheng G, Boyd A. Polarity effect on dichlorobenzene sorption by hexadecyltrimethylammonium-exchanged clays. Clays and Clay Minerals. 2000;**48**:43-50

[48] Lee S, Kim S. Study on the exchange reaction of HDTMA with the inorganic cations in reference montmorillonites. Geosciences Journal. 2003;**7**:203-208

[49] Hrachová J, Chodák I, Komadel P. Modification and characterization of montmorillonite fillers used in composites with vulcanized natural rubber. Chemical papers. 2008;**63**:55-61

[50] Sanchez-Martin MJ, Dorado MC, del Hoyo C, Rodriguez-Cruz MS. Influence of clay mineral structure and surfactant nature on the adsorption capacity of surfactants by clays. Journal of Hazardous Materials. 2008;**150**:115-123

[51] Polubesova T, Rytwo G, Nir S, Serban C, Margulies L. Adsorption of benzyltrimethylammonium and benzyltriethylammonium on

montmorillonite experimental studies and model calculations. Clays and Clay Minerals. 1996;**45**(6):834-841

[52] Shen Y. Phenol sorption by organoclays having different charge characteristics, colloids and surfaces a: Physicochem. Engineering Aspects. 2004;**232**:143-149

[53] Park Y, Ayoko G, Frost RL. Application of organoclays for the adsorption of recalcitrant organic molecules from aqueous media. Journal of Colloid and Interface Science. 2010; **354**:292-305

[54] Akcay G, Kılınc E, Akcay M. The equilibrium and kinetics studies of flurbiprofen adsorption onto tetrabutylammonium montmorillonite (TBAM). Colloids and Surfaces A. 2009; **335**:189-193

[55] Permien T, Lagaly G. The rheological and colloidal properties of bentonites dispersions in the presence of organic compounds V. Bentonites and sodium montmorillonite and surfactants, Clays and Clay Minerals. 1995;**43**:229-236

[56] Günister E, Alemdar SA, Güngör N. Effect of sodium dodecyl sulfate on flow and electrokinetic properties of Na-activated bentonite dispersions. Bulletin of Material Science. 2004;**27**(3):317-322

[57] Lin S, Teng M, Juang R. Adsorption of surfactants from water onto raw and HCl-activated clays in fixed beds. Desalination. 2009;**249**:116-122

[58] McLauchlin AR, Thomas NL. Preparation and characterization of organoclays based on an amphoteric surfactant. Journal of Colloid and Interface Science. 2008;**321**:39-43

[59] Zampori L, Gallo Stampino P, Cristiani C, Dotelli G, Cazzola P. Synthesis of organoclays using non-ionic surfactants: Effect of time, temperature and concentration. Applied Clay Science. 2010;**48**:97-102

[60] Hedley CB, Yuan G, Theng BKG. Thermal analysis of montmorillonites modified with quaternary phosphonium and ammonium surfactants. Applied Clay Science. 2007;**35**:180-188

[61] Patel HA, Somani S, Bajaj CH, Jasra RV. Preparation and characterization of phosphonium montmorillonite with enhanced thermal stability. Applied Clay Science. 2007;**35**:194-200

[62] Jeschke F, Meleshyn A. A Monte Carlo study of interlayer and surface structures of tetraphenylphosphonium-modified Na-montmorillonite. Geoderma. 2011;**169**:33-40

[63] Calderon J, Lennox B, Kamal MR. Thermally stable phosphonium-montmorillonite organoclays. Applied Clay Science. 2008;**40**:90-98

[64] Avalos F, Ortiz J, Zitzumbo R, López-Manchado M, Verdejo R, Arroyo M. Phosphonium salt intercalated montmorillonites. Applied Clay Science. 2009;**43**:27-32

[65] Li J, Zhua L, Cai W. Characteristics of organobentonite prepared by microwave as a sorbent to organic contaminants in water, colloids and surfaces a: Physicochem. Eng. Aspects. 2006;**281**:177-183

[66] Korichi S, Elias A, Mefti A. Characterization of smectite after acid activation with microwave irradiation. Applied Clay Science. 2009;**42**:432-438

[67] Gunawan NS, Indraswati N, Ju Y, Soetaredjo FE, Ayucitra A, Ismadji S. Bentonites modified with anionic and cationic surfactants for bleaching of

crude palm oil. Applied Clay Science. 2010;**47**:462-464

[68] Liu B, Wang X, Yang B, Sun R. Rapid modification of montmorillonite with novel cationic Gemini surfactantsand its adsorption for methyl orange. Materials Chemistry and Physics. 2011;**130**:220-1226

[69] Sullivan EJ, Hunter D, Bowman RS. Topological and thermal properties of surfactant-modified clinoptilolite studied by tapping-mode atomic force microscopy and high-resolution thermogravimetric analysis. Clays and Clay Minerals. 1997;**45**:42-53

[70] Giles C, Macewan TH, Nalkhwa SN, Smith D. Studies in adsorption. XI. A system of classification of solution adsorption isotherms. Journal of Chemical Society. 1960:3973-3993

[71] McAtee JL. Determination of random interstratification in montmorillonite. The American Mineralogist. 1956;**41**:627-631

[72] Rowland RA, Weisse EJ. Bentonite-methylamine complexes. Clays and Clay Minerals. 1963;**10**:460-468

[73] Lagaly G, Weiss A. Determination of the layer charge in mica-type layer silicates. In: Proceedings of the International Clay Conference, Toky. 1969. pp. 61-80

[74] Gherardi B. Organisation multiéchelle et stabilité colloïdale de suspensions d'argile organophile en milieu organique. Thèse Université d'Orléans; 1998

[75] Vaia RA, Teukolsky RK, Giannelis EP. Interlayer structure and molecular environment of alkylammonium layered silicates. Chemistry of Materials. 1994;**6**: 1017-1022

[76] Hackett E, Manias E, Giannelis EP. Molecular dynamics simulations of organically modified layered silicates. Journal of Chemical Physics. 1998;**108**: 7410-7415

[77] Ghiaci M, Abbaspur A, Kiaa R, Seyedeyn-Azad F. Equilibrium isotherm studies for the sorption of benzene, toluene, and phenol onto organo-zeolites and as-synthesized MCM-4. Separation and Purification Technology. 2004;**40**: 217-229

[78] Zhou Q, Frost LR, He H, Xi Y. Changes in the surfaces of adsorbed Para-nitrophenol on HDTMA organoclay—The XRD and TG study. Journal of Colloid and Interface Science. 2007;**307**:50-55

[79] Nahhal JY, Undabeytia T, Polubesova T, Mishael YG, Nir S, Rubin B. Organo-clay formulations of pesticides: Reduced leaching and photodegradation. Applied Clay Science. 2001;**18**:309-326

[80] Carrizosa MJ, Koskinen WC, Hermosin MC, Cornejo J. Dicamba adsorption-desorption on organoclays. Applied Clay Science. 2001;**18**:223-231

[81] Elguendi MS. Adsorption kinetics of cationic dyes stuffs onto natural clay. Adsorption Science and Technology. 1995;**13**:295-303

[82] Choi YS, Cho JH. Color removal dyes from wastewater using vermiculite. Environmental Technology. 1996;**17**: 1169-1180

[83] Özan SA, Erdem B, Özcan A. Adsorption of acid blue 193 from aqueous solutions onto Na–bentonite and DTMA–bentonite. Journal of Colloid and Interface Science. 2004;**280**: 44-54

[84] Ma J, Cui B, Dai J, Li D. Mechanism of adsorption of anionic dye from aqueous solutions onto organobentonite. Journal of Hazardous Materials. 2011; **186**:1758-1765

[85] Anirudhan TS, Ramachandran M. Adsorptive removal of tannin from aqueo Ramachandran us solutions by cationic surfactant-modified bentonite clay. Journal of Colloid Interface Science. 2015;**299**:116-124

[86] Vimonses V. Kinetic study and equilibrium isotherm analysis of Congo red adsorption by clay materials. Chemical Engineering Journal. 2009;**148**: 354-364

[87] Tahir H, Hammed U, Sultan M, Jahanze Q. Batch adsorption technique for the removal of malachite green and fast green dyes by using montmorillonite clay as adsorbent. African Journal of Biotechnology. 2010:8206-8214

[88] Ho YS, Chiang CC, Hsu YC. Sorption kinetics for dye removal from aqueous solution using activated clay. Journal Separation Science and Technology. 2001;**36**:2473-2488

[89] Oyanedel-Craver V, Smith A. Effect of quaternary ammonium cation loading and pH on heavy metal sorption to Ca bentonite and two organobentonites. Journal of Hazardous Materials. 2006; **137**:1102-1114

[90] Oyanedel-Craver VA, Megan F, Smith JA. Simultaneous sorption of benzene and heavy metals onto two organoclays. Journal of Colloid and Interface Sciences. 2007;**309**:485-492

[91] Guimarães A, Ciminelli V, Vasconcelos W. Smectite organofunctionalized with thiol groups for adsorption of heavy metal ions. Applied Clay Science. 2009;**42**:410-414

[92] Jiang J, Cooper C, Ouki S. Comparison of modified montmorillonite adsorbents part I: Preparation, characterization and phenol adsorption. Chemosphere. 2002;**47**: 711-716

[93] Szanto F, Dekany I, Patzk A, Varkonyi B. Wetting, swelling and sediment volumes of organophilic clays. Colloids and Surfaces. 1986;**18**:359-371

[94] Tvardovski AV, Fomkin AA, Tarasevich YI, Zhukova AI. Adsorptive deformation of Organo-substituted laminar silicates. Journal of Colloid and Interface Science. 1999;**212**:426-430

[95] Bors J, St D, Riebe B. Retention of radionuclides by organophilic bentonite. Engineering Geology. 1999;**54**:195-206

[96] Behnsen J, Riebe B. Anion selectivity of organobentonites. Applied Geochemistry. 2008;**23**:2746-2752

[97] Yah WO, Takahara A, Lvov Y. Selective modification of Halloysite lumen with Octadecyl Phosphonic acid: New inorganic tubular micelle. Journal of the American Chemical Society. 2012; **134**:1853-1859

[98] Owoseni O, Nyankson E, Zhang Y, Adams SJ, He J, McPherson GL, et al. Release of surfactant cargo from Interfacially-active Halloysite clay nanotubes for oil spill remediation. Langmuir. 2014;**30**:13533-13541

[99] Yapar S, Zbudak VO, Dias A, Lopes A. Effect of adsorbent concentration to the adsorption of phenol on hexadecyl trimethyl ammonium-bentonite. Journal of Hazardous Materials. 2005;**121**: 135-139

[100] Mojović Z, Jović-Jovičić N, Banković P, Žunić M, Abu R-SA, Milutinović-Nikolić A, et al.

Electrooxidation of phenol on different organo bentonite-based electrodes. Applied Clay Science. 2011;**53**:331-335

[101] Boyd SA, Shaobai S, Lee J, Mortland MM. Pentachlorophnol sorption by organo-clays. Clays and Clay Minerals. 1988;**36**:125-130

[102] Stapleton MG, Sparks DL, Dentel SK. Sorption of pentachlorophenol to HDTMA-clay as a function of ionic strength and pH. Environmental Science Technology. 1994;**28**:2330-2335

[103] Mortland MM, Shaobais S, Boyd S. Clay-organic complex as adsorbants for phenol and chlorophenol. Clays and Clay Minerals. 1986;**34**:581-585

[104] Zaghouane-Boudiaf H, Boutahala M. Preparation and characterization of organo-montmorillonites. Application in adsorption of the 2,4,5-trichlorophenol from aqueous solution. Advanced Powder Technology. 2011;**22**:735-740

[105] Bartelt-Hunt L, Susan E, Burns SJA. Nonionic organic solute sorption onto two organobentonites as a function of organi-carbon content. Journal of Colloid Interface Science. 2003;**266**:251-258

[106] Borisover M, Bukhanovsky N, Lapides I, Yariv S. Thermal treatment of organoclays: Effect on the aqueous sorption of nitrobenzene on n-hexadecyltrimethyl ammonium montmorillonite. Applied Surface Science. 2010;**256**:5539-5544

[107] Celis R, Koskinen WC, Hermosın MC, Ulibarri MA, Cornejo J. Triadimefon interactions with Organoclays and Organohydrotalcites. Soil Science Society America Journal. 2000;**64**:36-43

[108] Mishael G, Undabeytia T, Rytwo G, Sternberg BP, NIR S. Sulfometuron incorporation in cationic micelles adsorbed on montmorillonite. Journal of Agriculture Food Chemistry. 2002;**50**: 2856-2863

[109] Yaron-Marcovich D, Shlomo N, Chen Y. Fluridone adsorption–desorption on organo-clays. Applied Clay Science. 2004;**24**:167-175

[110] Galán-Pérez JA, Gámiz B, Celis R. Granulated organoclay as a sorbent to protect the allelochemical scopoletin from rapid biodegradation in soil. Environmental Technology & Innovation. 2022;**28**:102707

[111] Saitoh T, Shibayama T. Removal and degradation of -lactam antibiotics in water using didodecyldimethylammonium bromide-modified montmorillonite organoclay. Journal of Hazardous Materials. 2016;**317**:677-685

[112] Maia GS, Andrade JR, Silva MGC, Vieira MGA. Adsorption of diclofenac sodium onto commercial organoclay: Kinetic, equilibrium and thermodynamic study. Powder Technology. 2019;**345**: 140-150

[113] De Oliveira T, Boussafir M, Fougère L, Destandau E, Sugahara Y, Guégan R. Use of a clay mineral and its nonionic and cationic organoclay derivatives for the removal of pharmaceuticals from rural wastewater effluents. Chemosphere. 2020;**259**: 127480

[114] Imanipoor J, Mohammadi M, Mohammad D. Evaluating the performance of L-methionine modified montmorillonite K10 and 3-aminopropyltriethoxysilane functionalized magnesium phyllosilicate organoclays for adsorptive removal of azithromycin from water. Separation and Purification Technology. 2021;**275**: 119256

[115] Shen D, Jianxin F, Weizhi Z, Baoyu G, Qinyan Y, Kang Q. Adsorption kinetics and isotherm of anionic dyes onto organo-bentonite from single and multisolute systems. Journal of Hazardous Materials. 2009;**179**:99-107

[116] Gamoudi G, Srasra E. Adsorption of organic dyes by HDPy-modified clay: Effect of molecular structure on the adsorption. Journal of Molecular Structure. 2019;**1193**:522-531

[117] Belhouchat H, Zaghouane-Boudiaf H, Viseras C. Removal of anionic and cationic dyes from aqueous solution with activated organo-bentonite/sodium alginate encapsulated beads. Applied Clay Science. 2017;**2017**:9-15

[118] Lee SH, Song DI, Jeon YW. An investigation of the adsorption of organic dyes onto organo-montmorillonite. Environmental Technology. 2001;**22**:247-254

[119] Gamoudi G, Srasra E. Removal of cationic and anionic dyes using purified and surfactant-modified Tunisian clays: Kinetic, isotherm, thermodynamic and adsorption-mechanism studies. Clay Minerals. 2018;**53**:1-16

[120] Santoso SP, Angkawijaya AE, Yuliana M, Bundjaja V, Soetaredjo FE, Ismadji S, et al. Saponin-intercalated organoclays for adsorptive removal of β-carotene: Equilibrium, reusability, and phytotoxicity assessment. Journal of the Taiwan Institute of Chemical Engineers. 2020;**117**:198-208

[121] Bors J, Dultz S, Riebe B. Organophilic bentonites as adsorbents for radionuclides I. Adsorption of ionic fission products. Applied Clay Science. 2000;**16**:1-13

[122] Dultz S, Riebe B, Bunnenberg C. Temperature effects on iodine adsorption on organo-clay minerals II. Structural effects. Applied Clay Science. 2005;**28**:17-30

[123] Krishna BS, Murty DSR, Prakash BS. Surfactant-modified clay as adsorbent for chromate. Applied Clay Science. 2001;**20**:65-71

[124] Brum MC, Capitaneo JL, Oliveira JF. Removal of hexavalent chromium from water by adsorption onto surfactant modified montmorillonite. Minerals Engineering. 2010;**23**:270-272

[125] Xi Y, Mallavarapu M, Naidu R. Preparation, characterization of surfactants modified clay minerals and nitrate adsorption. Applied Clay Science. 2010;**48**:92-96

[126] Gammoudi S, Frini-Srasra N, Srasra E. Nitrate sorption by organosmectites. Engineering Geology. 2012a;**124**:119-129

[127] Gammoudi S, Frini-Srasra N, Srasra E. Kinetic and equilibrium studies of fluoride sorption onto surfactant-modified smectites. Clay Minerals. 2012b;**47**:429-440

[128] Li D, Kaplan D, Sams A, Powell BA, Knox AS. Removal capacity and chemical speciation of groundwater iodide (I^-) and iodate (IO_3^-) sequestered by organoclays and granular activated carbon. Journal of Environmental Radioactivity. 2018;**192**: 505-512

[129] Bagherifam S, Brown TC, Fellows CM, Naidu B, Komarneni S. Highly efficient removal of antimonite (Sb (III)) from aqueous solutions by organoclay and organozeolite: Kinetics and isotherms. Applied Clay Science. 2021;**203**:106004

[130] Volzone C, Rinaldi JO, Ortiga J. Retention of gases by

hexadecyltrimethyl-ammonium montmorillonite clays. Journal of Environmental Management. 2006;**79**: 247-252

[131] Volzone C. Retention of pollutant gases: Comparison between clay minerals and their modified products. Applied Clay Science. 2007;**36**:191-196

[132] Kinjal JS, Toyoko I, Masaki U, Shing-Jong H, Pei-Hao W, Liu S. Poly (amido amine) dendrimer-incorporated organoclays as efficient adsorbents for capture of NH_3 and CO_2. Chemical Engineering Journal. 2017;**312**:118-125

[133] Park S, Ryu J, Cho YH, Sohn D. Halloysite nanotubes loaded with HKUST-1 for CO_2 adsorption. Colloids and Surfaces A: Physicochemical and Engineering Aspects. 2022;**651**:129750

[134] Herrera P, Burghard RC, Phillips TD. Adsorption of salmonella enteritidis by cetylpyridinium-exchanged montmorillonite clays. Veterinary Microbiology. 2000;**74**:259-272

[135] Herrera P, Burghardt R, Huebner HJ, Phillips TD. The efficacy of sand-immobilized organoclays as filtration bed materials for bacteria. Food Microbiology. 2004;**21**:1-10

[136] Ma YL, Yang B, Guo T, Xie L. Antibacterial mechanism of Cu2+– ZnO/ cetylpyridinium–montmorillonite in vitro. Applied Clay Science. 2010;**50**: 348-353

[137] Bujdáková H, Bujdáková V, Májeková-Koščová H, Gaálová B, Bizovská V, Boháč P, et al. Antimicrobial activity of organoclays based on quaternary alkylammonium and alkylphosphonium surfactants and montmorillonite,d. Applied Clay Science. 2018;**158**:21-28

[138] Abhinaya R, Mangalara D. Effect of surfactant modified Halloysite nanotube on growth and biofilm formation of gram-positive bacteria. Materials Today: Proceedings. 2019;**18**:1709-1715

Chapter 5

Chemical Enhanced Oil Recovery Using Ionic Liquid-Based Surfactants

Mona Kharazi and Javad Saien

Abstract

Critical challenges have forced the oil industry to improve chemical enhanced oil recovery (CEOR) processes using more effective materials. In this regard, ionic liquid-based surfactants (IL-based surfactants) with noteworthy features such as good interfacial activity, recyclability, environmentally friendliness, and stability at high temperatures and salinity can be considered as a significant option. While IL-based surfactants have been of interest in various fields, their applications in CEOR have not been systematically reviewed. In this frame, this chapter overviews the performance of IL-based surfactants in different portions of oil production. For this purpose, a brief explanation of the characteristics of the IL-based surfactants and their applicability in CEOR is first provided. Then, the performance of the IL-based surfactants in surfactant flooding, micellar flooding, and wettability alteration are surveyed in detail. Also, the activity of IL-based surfactants in all three regions has been monitored with the effect of the chemical structure, as well as comparing the performance of different types of IL-based surfactants. Relevantly, a comparison with conventional surfactants is also included. The crucial challenges and possible directions for the use of IL-based surfactants and the conclusions are summarized in the final section.

Keywords: ionic liquid-based surfactants, surfactant flooding, micellar flooding, wettability alteration, chemical enhanced oil recovery

1. Introduction

With existing many challenges facing the oil industry, including dwindling crude oil reserves and the limited possibility of finding new oil fields, and more importantly, only one-third of the original oil in place (OOIP) can be recovered through primary and secondary oil recovery, the oil industry needs to pursue chemical enhanced oil recovery (CEOR) techniques [1]. Among the various methods, the use of surfactants, which focuses on reducing interfacial tension (IFT) and microemulsion formation as well as wettability modification has attracted much attention [2]. However, the low resistance and poor performance of conventional surfactants at high temperatures and salinity conditions have convinced researchers to use more effective and economical materials in this field [3].

IntechOpen

Ionic liquid-based surfactants (IL-based surfactants) can be considered a significant alternative in CEOR. Most IL-based surfactants exhibit powerful interfacial activity compared to conventional surfactants with the same molecular structure [4, 5]. IL-based surfactants by adsorbing at the oil-water interface can reduce IFT and form microemulsion, and by changing the properties of reservoir rock surfaces alter wettability, which completely rises the mobility of crude oil in the reservoirs, and as a result of oil production increases [3, 5]. Relevantly, the unique properties of IL-based surfactants, such as stability at high temperatures and salinity as well as their environmentally friendly nature, make them suitable for CEOR [6, 7]. As another effect, high molecular weight IL-based surfactants can increase the viscosity of the injected fluid and provide sweeping efficiency [8]. Additionally, IL-based surfactant molecules are adjustable and by inserting the appropriate cation/anion into their structures, task-specific IL-based surfactants can be employed in CEOR [4]. Accordingly, various types of IL-based surfactants, including Cationic, Anionic, Multicationic, Catanionic, and Zwitterionic, have been synthesized and their remarkable properties have been investigated [3].

Despite IL-based surfactants indicating favorable results in the oil industry, screening their performance in different aspects of oil recovery is a necessary step before using them for large-scale CEOR. In spite of many efforts, no comprehensive report has been provided in this field. Thus, this chapter first provides a brief description of the IL-based surfactant characteristics and their applicability in the oil industry. In the next step, the performance of the IL-based surfactant in the important regions of CEOR is surveyed. These regions include the following:

- Surfactant flooding
- Micellar flooding
- Wettability alteration

Relevantly, the effect of the chemical structure of the IL-based surfactant and the characteristics and outstanding performance of the new types of IL-based surfactant are considered to highlight the advantages of the respective materials [3, 4, 7, 9]. To select the most suitable surfactants and display higher interfacial activity, a brief comparison between the IL-based surfactant and conventional surfactants is presented subsequently. In the final step, the important challenges as well as possible directions and conclusions of this chapter are summarized in the last section. Overall, this chapter aims is to provide in-depth information on the use of IL-based surfactants in the oil industry and to pave the way for the development of better IL-based surfactants that may meet the requirements of the CEOR processes.

2. Characteristics of ionic liquids

2.1 Ionic liquids and their properties

Ionic liquids (ILs) are composed of organic cations and inorganic/organic anions [4]. The steric asymmetry between the size and morphology of the ions causes mismatches and the formation of unstable crystal lattices, which lowers the melting point [10]. ILs have melting points below 100°C, and room temperature ionic liquids

(RTILs) specifically refer to salts that form electrolytes that are liquid in the room temperature range or even below [3]. For a better explanation, a comparison between the structure of the IL and common salt is presented in **Figure 1**.

The remarkable properties of IL, such as high chemical and thermal stability even under harsh conditions, low vapor pressure, recyclability, non-flammability, and an environmentally friendly manner, confirm a large number of applications of these materials [11, 12]. Furthermore, the properties of IL can be upgraded by modifying/altering various parts of their molecular structure, usually by simple strategies, making them "designer" materials [13, 14]. In this regard, many organic cations can be coupled to various inorganic or organic anions, ranging from simple inorganic ions to more complex organic species. For better illustration, some commonly used cations and anions are presented in **Figure 2**. As well, some general ILs specifications are listed in **Table 1**.

2.2 Ionic liquid-based surfactant

IL-based surfactants are a class of ILs that have both hydrophilic (lipophobic, water liking, or oil disliking) and hydrophobic (lipophilic, oil liking, or water

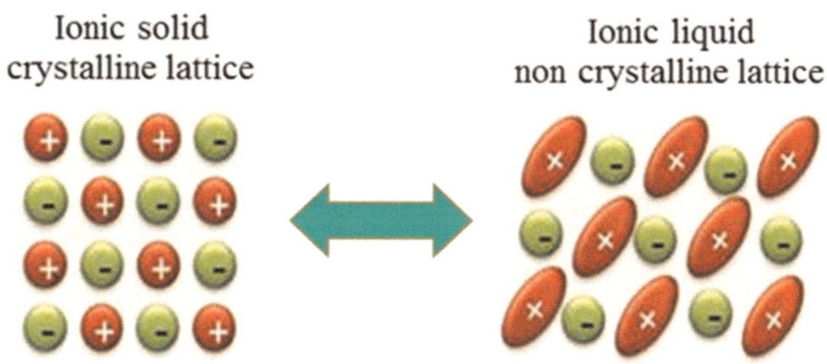

Figure 1.
Comparison between the structure of the IL and common salt.

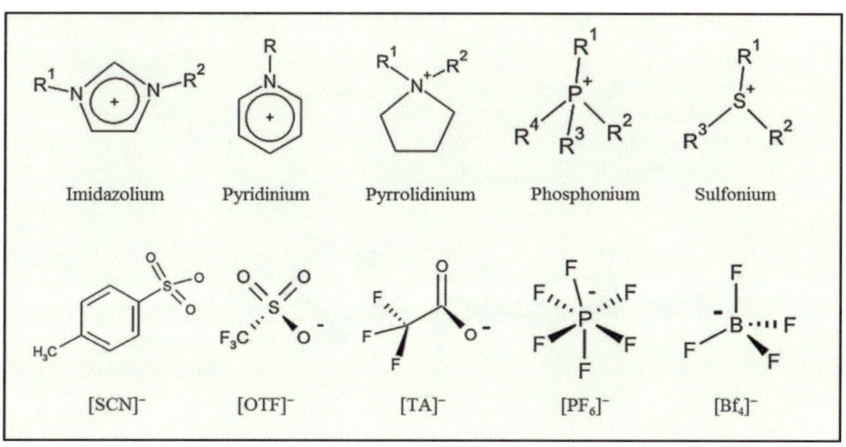

Figure 2.
Some conventional IL cations and anions.

Melting point	Below 100°C
Liquidus range	Often more than 200°C
Thermal stability	High
Chemical stability	High
Viscosity	Normally less than 100 cPo
Polarity	Moderate
Specific conductivity	Less than 10 $mS.cm^{-1}$
Solvent or catalyst	Good for a lot of reactions
Vapor pressure	Negligible

Table 1.
General ILs specifications [3].

disliking) groups that are "amphiphilic" in nature [3, 4]. To better illustrate, the three major parts of the chemical structure of [C_6mim][Br], as a common Cationic IL-based surfactant, include the hydrophobic side chains and the hydrophilic head group, which form the cationic portion, as well as the counter anion are shown in **Figure 3**.

The ionic or polar hydrophilic group gives water solubility to IL-based surfactants; however, the organic hydrophobic group gives them oil solubility. After the dissolving of IL-based surfactants in the aqueous or oil phase, like conventional surfactants, their molecules are adsorbed at the oil-water interface in the way that the hydrophilic portions are adjacent to the water phase and the hydrophobic portions are adjacent to the oil phase. This structure provides different interfacial activities for the IL-based surfactant such as IFT reduction and emulsion formation. On the other hand, partial adsorption of IL-based surfactant on the solid surface of rocks changes the surface properties, and depending on the interface orientation the surface wettability can be changed [5]. These points will be discussed in detail in the following sections.

2.3 Impact of the structure of ionic liquid-based surfactant

There is a strong relationship between the chemical structures of IL-based surfactant and their properties. In general, IL-based surfactant has three major parts in its

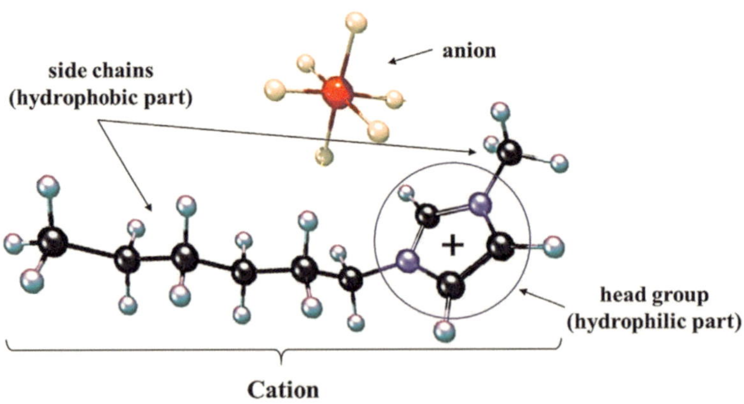

Figure 3.
The three major parts of the chemical structure of [C_6mim][Br].

molecule: alkyl chain, head group, and counter anion. By changing the structure of these parts, the characteristics of the IL-based surfactant can be modified for specific applications [15, 16]. For example, the hydrophobicity, interfacial activity, miscibility, and thermal stability of IL-based surfactant can be improved by altering the structure of the counter anions, while the hydrophobicity, interfacial activity, adsorption tendency, aggregate formation, viscosity, and density are significantly changed with head groups and/or the alkyl chains structure [3]. In this regard, an overview of the characteristics of IL-based surfactants concerning to different parts of their structure is described here.

2.3.1 Alkyl chain

The length of alkyl chains and the number of involved branches are two important factors for IL-based surfactant characteristics that affect the hydrophobicity, solubility, interfacial activity, and adsorption tendency of IL-based surfactants. A longer alkyl chain provides more adsorption tendency, hydrophobicity, and interfacial activity for IL-based surfactant, however, the solubility in the water phase is reduced. The longer hydrophobic chain also causes more van der Waals attraction between the IL-based surfactants molecules and the crude oil phase, leading to highly stable adsorption at the interface. The same behavior has been observed for the second alkyl chain in the IL-based surfactant structure, if present [17, 18].

2.3.2 Head group

The structure of the head group can affect the hydrophilicity character of the IL-based surfactant. The size of the head groups, the presence of the π-electron ring, and their polarity have distinct effects on the hydrophilicity and interfacial activity of IL-based surfactants. Generally, due to their greater hydrophobicity, poor hydration, and low electrostatic repulsion forces, bulky heterocyclic head groups with π-electron rings cause better IFT reductions compared to those with simple groups. These head groups facilitate migration and lead to greater adsorption at the interface. In this respect, the widely studied IL-based surfactant for IFT reduction includes bulky and asymmetric organic cations such as imidazolium, pyridinium, ammonium, sulfonium, phosphonium, pyrrolidinium, pyrazolium, thiazolium, and guanidinium [19, 20].

2.3.3 Counter anion

The nature of the counter anions affects the miscibility of IL-based surfactants in polar/non-polar solvents and their interfacial activity, although to a lesser extent compared to the alkyl chain and the head group. The molecular dynamics simulations studies confirm that bulkier and more polarizable counter anions increase the interfacial activity of IL-based surfactants. This is because large counter anions are less prone to effective hydration, making them easier to adsorb at the interface, which reduces the electrostatic repulsion between the head groups and facilitates IFT reduction. Also, cations can easily penetrate the electron cloud of the bulky and polarizable counter anions, which facilitates interfacial adsorption. Some common inorganic and organic anions in IL-based surfactants are hexafluorophosphate $[PF_6]^-$, tetrafluoroborate $[BF_4]^-$, acetate, $[Ac]^-$, alkyl sulfate $[AS]^-$, bis(trifluoromethylsulfonyl)imide $[NTf_2]^-$, trifluoromethanesulfonate, $[OTF]^-$, and halides [13, 21].

In some IL-based surfactants, the counter anions have an alkyl chain that affects the activity of IL-based surfactant. In these cases, by increasing alkyl chain length, more hydrophobicity and less hydration are expected, leading to higher IL-based surfactants adsorption even at low concentrations, which provides the higher interfacial activity [22].

2.4 New types of IL-based surfactants

Various types of IL-based surfactants with specific properties and different structures have been used in the oil industry, which is schematically shown in **Figure 4**. In general, IL-based surfactants can be classified as follows based on the nature of the surface active moiety and the number of head groups [3, 4, 23].

The "Cationic" type, in which surface active parts have a positive charge and are the most usual IL-based surfactant.

The "Anionic" type, consists of long alkyl chain anions, with or without branches, coupled to rather large heterocyclic cations. As an important advantage, Anionic IL-based surfactants have lower toxicity compared to other types of IL-based surfactants.

The "Catanionic" type, in which both cationic and anionic parts have a surface active character. This structure provides the higher interfacial activity. The asymmetry in the cation and anion parts, as well as linear or branched alkyl chains, are the factors affecting their activity.

The "Multicationic" type, in which more than one hydrophilic head group, each attached to hydrophobic alkyl chains, is connected with a spacer giving them high interfacial activity. This structure provides good hydrophobicity, high IFT reduction, desired micellization, and low toxicity. Here, the length and structure of the spacer play an important role in the hydrophobicity and molecular orientation at the interface.

The "Zwitterionic" type, in which the cationic and anionic parts are linked together by a covalent bond. With this structure, the electrostatic repulsion of charged groups is neutralized giving ordered and intense molecular orientations at the interface.

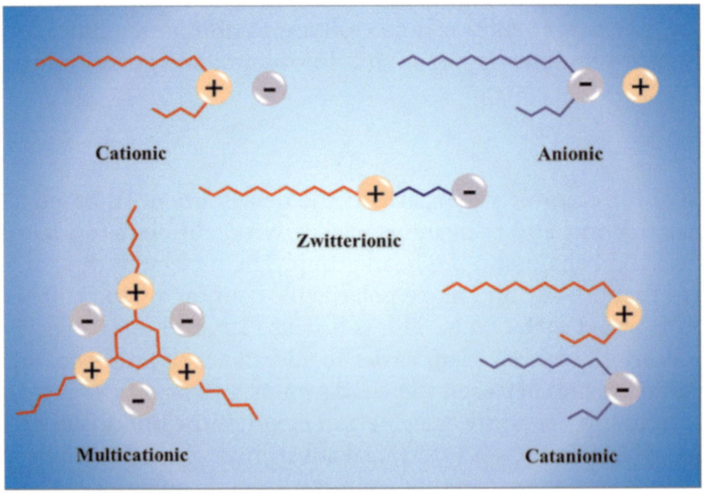

Figure 4.
Schematic of the different types of IL-based surfactants.

It is worth noting that novel nanostructured IL-based surfactants have recently been successfully designed, synthesized, and tested under various conditions [24–26]. This new type of IL-based surfactant shows good interfacial activity and exhibits some nanoparticle properties that make it significant material in various aspects of the oil industry. However, their investigation is in the early stages and more research is needed in this area.

3. Chemical enhanced oil recovery and ionic liquid-based surfactants

3.1 Enhanced oil recovery

Crude oil is a natural source of petroleum that consists of carbon, nitrogen, oxygen, and sulfur elements as well as traces of metals. Crude oil naturally forms in the earth's porous rock deposits, "oil reservoirs". Such a complex mixture of crude oil is divided into four components: Saturates (S), Aromatics (A), Resins (R), and Asphaltenes (A), which are abbreviated as SARA [27]. Different crude oils are classified based on their composition, i.e. SARA percentage and API degree.

Although the energy consumption is growing worldwide and is estimated to quadruple by the year 2100 [28], dwindling crude oil reserves, the low probability of discovering new oil fields, and meeting demand for crude oil pose a serious challenge to the oil industry. Furthermore, the use of primary and secondary recovery techniques could only extract one-third of the original oil in place (OOIP) [2]. To solve these problems, most of the mature oil fields in the world require enhanced oil recovery (EOR) techniques to maintain their production levels. EOR is the process of injecting gases, chemicals, and/or using thermal energy to recover excess crude oil [1]. The main subdivisions of the EOR methods are shown in **Figure 5**. Among them, chemical enhanced oil recovery (CEOR) is notable due to its efficiency and low cost handling. In this method, selected chemicals such as surfactants, alkalis, polymers, and organic solvents are used. To facilitate a more comprehensive comparison, the advantages and disadvantages of different EOR methods are summarized in **Table 2**.

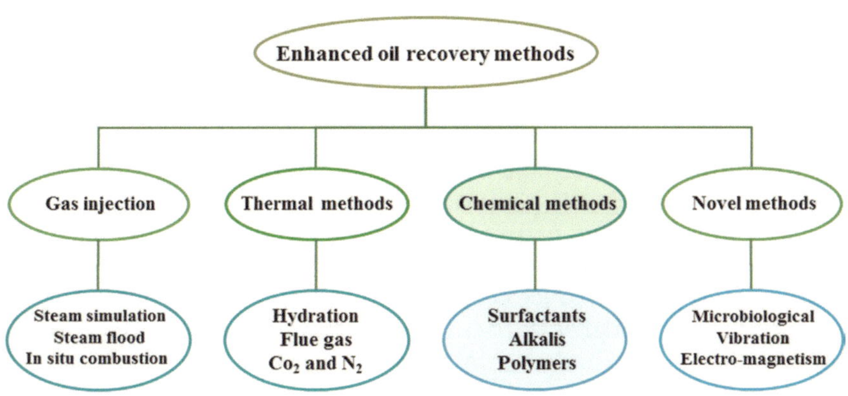

Figure 5.
The classification of EOR methods.

EOR methods	Advantages	Disadvantages
Gas injection	High efficiency Inexpensive	Gas corrosion Asphaltene problems
Thermal methods	Improved recovery factor Can be used for heavy crude oils	High cost Heat lost
Chemical methods	High efficiency No damage to reservoir rock	Sensitive to salinity and heat High cost
Novel methods	Environmental benefits Inexpensive in some cases	Low recovery in some case Sensitive to reservoirs condition

Table 2.
EOR methods: Advantages and disadvantages [1, 2].

The use of surfactants is mainly for surfactant flooding as well as micellar flooding [29]. These are the two major CEOR techniques that focus on reducing IFT and creating emulsions/microemulsions to increase the oil mobilization through porous rocks and direct residual oil flows toward the production well [5, 8]. The wettability alteration is another important mechanism by which surfactants can change the wettability of the reservoir rocks from oil-wet to water-wet [30]. With the change in wettability, the movement of crude oil in reservoirs smooths, and oil production increases [31]. Briefly, these three methods are the main applications of surfactants in CEOR.

3.2 Ionic liquid-based surfactants applicability in CEOR

Different types of conventional surfactants have been investigated for CEOR purposes. But almost all of them lose their effectiveness under harsh conditions of salinity and temperature [6]. For instance, The C-S-O bonds in sulfate-based surfactants are hydrolyzed at 60°C, and this causes the loss of performance of this group of surfactants above this temperature [32]. In this respect, IL-based surfactants can be considered a significant alternative. The stability of IL-based surfactants at high temperatures and salinities, as well as their notable amphiphilic nature, make them good candidates for CEOR. Based on this, IL-based surfactants reduce IFT and form emulsions by adsorbing and directing to the oil-water interfaces and by changing the properties of the rock surface to facilitate the wettability alteration.

Meanwhile, IL-based surfactants with high molecular weight increase the mobility of the crude oil by increasing the viscosity of the injected fluid, giving it a high sweeping effect. Due to their high thermal stability, IL-based surfactants are also suitable for thermal oil recovery and can enhance oil production by this method, particularly for heavy crude oils. Likewise, the aromaticity of some IL-based surfactants can assist oil production. As an additional benefit, IL-based surfactants may reduce water consumption during oil extraction, which in turn helps to separate sand and clay from crude oil and reduce corrosion problems.

Despite the high potential of IL-based surfactants for use in the CEOR process, the entire injected fluid system and reservoir conditions such as the nature of crude oil, rock type, and structure of IL-based surfactants must be considered for the operation. Consequently, the screening of the IL-based surfactants is a necessary step before using them for the CEOR process to correctly predict their behavior.

4. Surfactant flooding

High IFT values between crude oil and water are the main reason for the low mobility of residual crude oils in the reservoirs. IFT represents the extent of molecular cohesive forces at the interface of immiscible liquids, which arise from the unbalanced forces that molecules feel from adjacent molecules [33]. **Figure 6** shows a schematic diagram of the involved forces at the liquid-liquid interface. IFT plays a vital role in sketching and operating with immiscible fluids as well as the hydrodynamic and contact of the phases. Therefore, it is important to find a way to reduce IFT for higher oil production.

In the surfactant flooding method, IL-based surfactant molecules are adsorbed at the interface of crude oil-water. The adsorption of IL-based surfactant at the interface reduces the intermolecular forces between the phases and thus provides the IFT reduction. As the IFT decreases to an ultralow value, the capillary number (representing the relative effect of viscous drag forces versus IFT forces acting across an interface between crude oil-water in the reservoirs) increases [4]. As a result, the capillary forces that are the major cause of oil trapping in the pores are weakened and the production of crude oil from the reservoirs is significantly enhanced. The oil recovery mechanism based on IFT reduction is schematically presented in **Figure 7**.

Investigations show that using imidazolium IL-based surfactants, about 90% of the extra-heavy crude can be recovered from Utah and Canadian tar sands and IL-based surfactants could be recycled up to 5 times without significant performance loss even at high salinity levels [34].

4.1 Effect of ionic liquid-based surfactants structure

Because the decrease in IFT is attributed to the ability of IL-based surfactants to adsorb at the crude oil-water interface, slight changes in the structure of IL-based surfactants such as alkyl chain, head group, and counter anion, significantly alter the surfactant flooding process.

Accordingly, the longer the hydrophobic alkyl chain, the greater the ability of IL-based surfactants to reduce IFT. This can be attributed to the higher adsorption at the interface as well as the stronger van der Waals attraction between the alkyl chain of

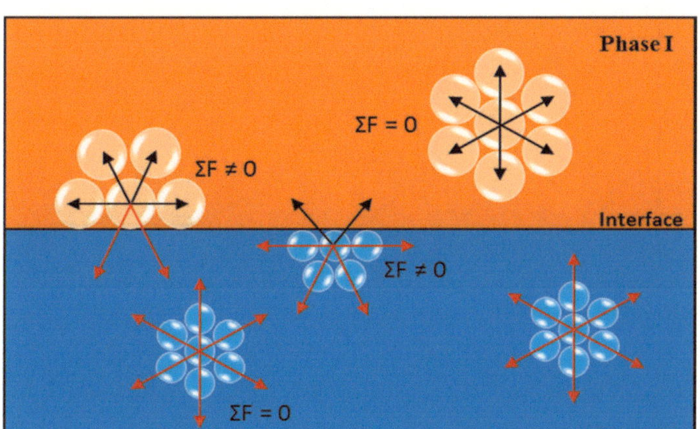

Figure 6.
Schematic of the involved forces at the liquid-liquid interface.

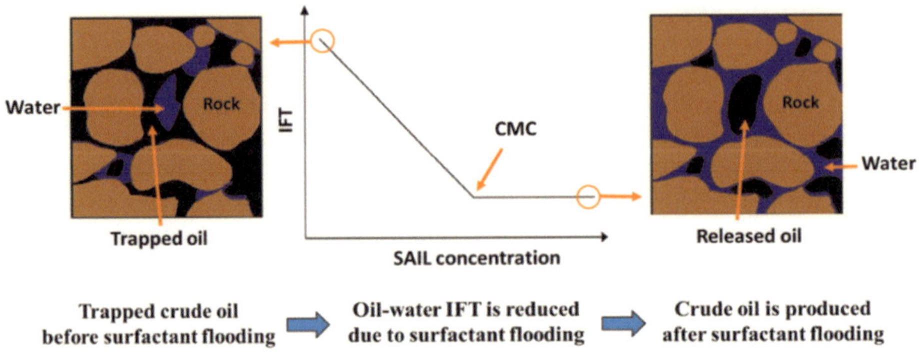

Figure 7.
Mechanism of crude oil recovery with the IFT reduction [5].

the IL-based surfactants and the hydrophobic part of the crude oil. Besides, branched alkyl chains meaningfully decrease the crude oil-water IFT. One advantage is that due to high performance, small amounts of long chain IL-based surfactants are required to achieve a certain IFT value that is economically and environmentally beneficial [3]. In this respect, comparing the effect of [C_8mim][Cl], [C_{12}mim][Cl], [C_8py][Cl], and [C_{12}py][Cl], indicates stronger performances with longer alkyl chain IL-based surfactants [18, 35, 36]. For more illustration, the change of IFT with the alkyl chain length of [$C_n$im-C_4-imC_n][Br_2] n = 4, 8, and 12 at different concentrations has been investigated in various researches [26, 37, 38]. The results of these studies are shown in **Figure 8**.

By investigating the effect of head groups, it was found that IL-based surfactants with bulky heterocyclic head groups as well as nonlocal charge rings such as imidazolium or thiazolium show higher interfacial activity and are greater effective in the IFT reduction. This phenomenon is due to the superior hydrophobicity of IL-based surfactants with heterocyclic head groups as well as the strong interaction of aromatic

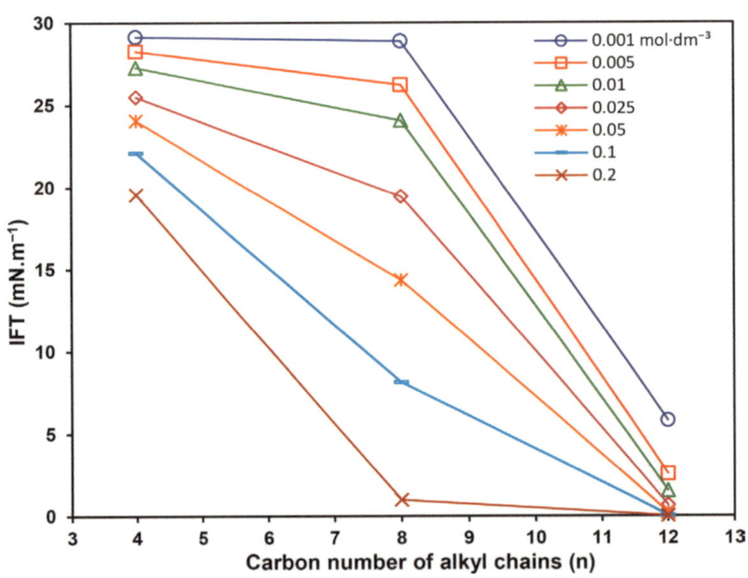

Figure 8.
IFT variation as a function of alkyl chain length for [$C_n$im-C_4-imC_n][Br_2] [26, 37, 38].

compounds of the crude oil with the aromatic head group of the IL-based surfactant. Likewise, the reduction of IFT by $[C_4mim][H_2PO_4]$ and $[C_6mim][HSO_4]$ with bulky heterocyclic imidazolium head groups was 37.8 and 42.4%; but for $[Et_2NH_2][H_2PO_4]$ and $[Et_3NH][HSO_4]$ with the quaternary ammonium head group is only 32.6 and 37.7%, respectively [39].

For counter anion, with increasing polarizability, more interface activity is obtained. This is attributed to the greater interaction of polarizable counter anions with the IL-based surfactant cations that reduce the repulsion between the cationic head groups at the interface and provide more reduction of IFT [40]. For example, using $[C_4mim][X]$ variants, the effectiveness was shown to be in the order $[H_2PO_4]^- > [PF_6]^- > [BF_4]^- > [HSO_4]^-$ [39]. The positive effect of the alkyl chains in counter anions is also attractive. For instance, for the applied cations of $[CP]^+$, $[BT]^+$, $[PA]^+$, and $[3\text{-}HPA]^+$, the IFT reduction was effective in the order $[C_6H_{13}COO]^- > [CH_3COO]^- > [HCOO]^-$ [22]. This is due to the interaction of alkyl chains and crude oil, which enhances the adsorption of IL-based surfactants at the interface.

4.2 New types of ionic liquid-based surfactants

New types of IL-based surfactants indicate better interfacial activity compared to conventional Cationic IL-based surfactants. In this regard, ultra-low IFT values can be obtained with Catanionic, Zwitterionic, and Multicationic IL-based surfactants. Notably, it is common to use co-surfactants to achieve ultra-low IFTs, but no co-surfactants have been used for these types [3]. These advantages reflect the stronger amphiphilic nature of the mentioned IL-based surfactants, which is the result of less electrostatic repulsion of charged moieties and closer molecular orientation at the interface, as well as multiple alkyl chains in their structure. Accordingly, the use of [P144,4][DS], as a Catanionic IL-based surfactant, has resulted in a reduction of the crude oil-water IFT to 0.1 $mN.m^{-1}$, which is much lower than about 10 $mN.m^{-1}$ that can be achieved by $[C_{12}py][Cl]$, $[C_8mim][Cl]$, and $[C_{12}mim][Cl]$, with the same concentrations [41].

In the case of the Zwitterionic IL-based surfactants, the covalent bonding between the cationic and anionic groups of the molecules neutralizes the electrostatic repulsion of the charged moieties and makes intense molecular arrangements at the interface. Correspondingly, comparing the interfacial activity of $[C_{16}IPS]$ and $[C_{16}IPS\text{-}Nsa]$ Zwitterionic IL-based surfactants with the same alkyl chains Cationic IL-based surfactants of $[C_{16}mpyr][Br]$ and $[C_{16}mim][Br]$, displays that $[C_{16}IPS]$ and $[C_{16}IPS\text{-}Nsa]$ can reduce the IFT from 20 to 1 and 0.1 $mN.m^{-1}$, respectively. However, the decrease was 5.8 and 3.7 $mN.m^{-1}$ for $[C_{16}mpyr][Br]$ and $[C_{16}mim][Br]$ respectively, under same concentrations [42].

As an important feature, ultra-low IFTs can be obtained using Multicationic IL-based surfactants. In this regard, Gemini IL-based surfactants, as a type of Multicationic IL-based surfactants, show that can significantly reduce IFT, even at low concentrations. A comparison between the performance of $[C_{12}im\text{-}C_4\text{-}imC_{12}][Br_2]$ and $[C_8im\text{-}C_4\text{-}imC_8][Br_2]$ Gemini IL-based surfactants with $[C_{12}mim][Cl]$ and $[C_8mim][Cl]$ as an analog Cationic IL-based surfactants demonstrate the better performance of Gemini IL-based surfactants [26, 37, 38]. Also, the effect of Gemini IL-based surfactants increases with extending the alkyl chain length. For instance, the long alkyl chain $[C_{12}im\text{-}C_4\text{-}imC_{12}][Br_2]$ was able to decrease the crude oil-water IFT from 32 to about 10^{-3} $mN.m^{-1}$ [38].

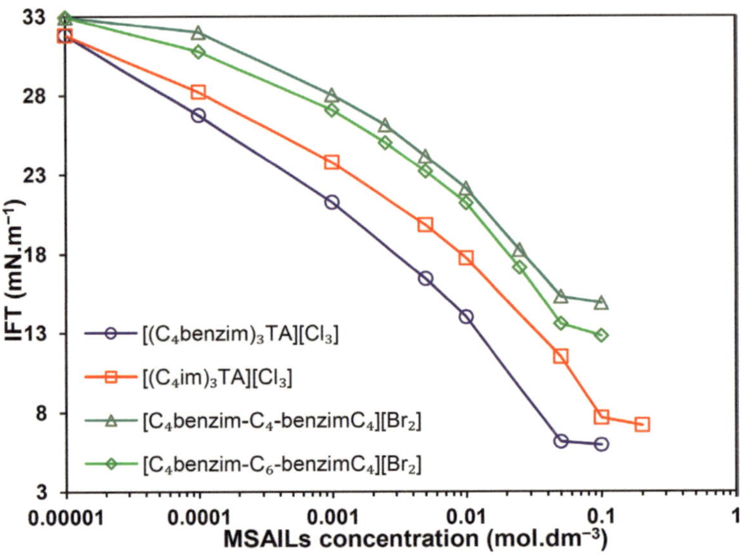

Figure 9.
Comparing IFT variation of a Tripodal IL-based surfactants and analog Gemini IL-based surfactants in crude oil-water system [43].

Tripodal IL-based surfactants, as the newest synthetized Multicationic IL-based surfactants, exhibit a noteworthy capability. For a better explanation, the performance of the benzimidazolium and imidazolium Tripodal IL-based surfactants with short chain lengths, namely $[(C_4benzim)_3TA][Cl_3]$ and $[(C_4im)_3TA][Cl_3]$, were compared with two structurally analog benzimidazolium Gemini IL-based surfactants $[C_4benzim\text{-}C_n\text{-}benzimC_4][Br_2]$ n = 4 and 6 [24, 25, 43]. The results are presented in **Figure 9**. As can be seen, the Tripodal IL-based surfactants produce a significant reduction in IFT even at lower concentrations that are economically important in CEOR. These results are due to greater hydrophobicity and the higher tendency to migrate toward the interface. Also, in comparison between the Tripodal IL-based surfactants, the benzimidazolium outperformed the imidazolium with 7.5% higher IFT reduction because of the better spread electrostatic repulsion between the positively charged head groups [43].

According to these results, new types of IL-based surfactants can provide the next generation of surfactants with distinct properties, especially in obtaining ultra-low IFT values.

5. Micellar flooding

In addition to "surfactant flooding", IL-based surfactants can solubilize crude oils in the core of micelles and form emulsion or microemulsion in a process called "micellar flooding" [8]. The emulsion is the dispersion of one phase in another immiscible phase in the form of small droplets. However, microemulsions are practically more effective in CEOR. Danielsson and Lindman [44], defined a microemulsion as "a system of water, oil, and an amphiphilic compound that has a single optically isotropic and thermodynamically stable liquid solution". The droplet size of microemulsions is smaller than conventional emulsions but bigger than micelles. Based on the Winsor organization, four types of microemulsions are predictable. Type I is an oil-in-water (O/W) microemulsion in which the oil is dissolved as micelles in a continuous aqueous water phase with an additional oil phase (Winsor I). Type II is a water-in-oil (W/O) microemulsion in which water is dissolved as reverse micelles in an oil phase with an additional water phase (Winsor II). Type III is a three-phase system where the intermediate phase microemulsion is balanced with additional oil and water phases (Winsor III). Type IV is a single-phase microemulsion in which oil and water phases are combined in one phase (Winsor IV) [8]. For a better explanation, four types of Winsor microemulsion structures are shown schematically in **Figure 10**.

For droplet formation and dissolution of O/W or W/O microemulsions, the interface must be divided into new small parts with more total interface areas, which requires a reduction in IFT. Therefore, to perform CEOR by micellar flooding, IL-based surfactants should decrease the crude oil-water IFT to about 10^{-3} mN.m^{-1} and it is essential that the IL-based surfactant concentration was higher than the corresponding CMC [3]. Therefore, the use of IL-based surfactants with low inherent CMC is more suitable [8]. To form O/W emulsions, the polar part of IL-based surfactants remains in the aqueous phase and the non-polar part extends into the oil phase. While for W/O emulsion, this arrangement is reversed. For better illustration, the formation of O/W and W/O emulsions with IL-based surfactants is presented in **Figure 11**.

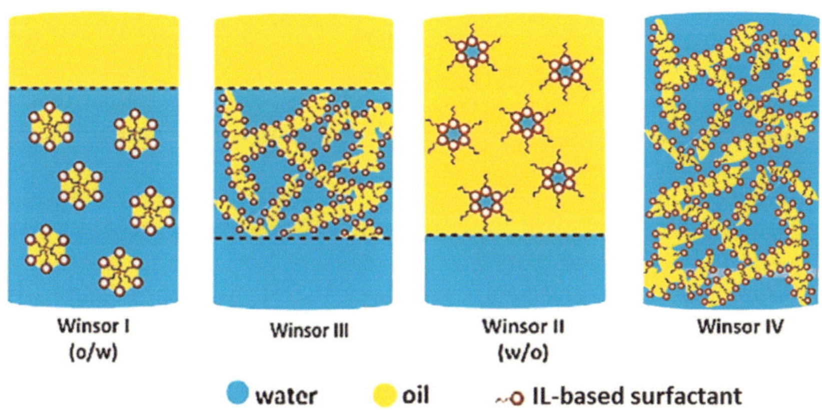

Figure 10.
Four types of Winsor microemulsion structures [8].

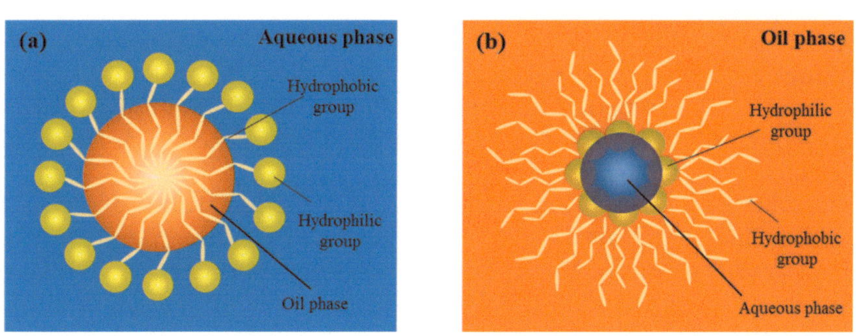

Figure 11.
Formation of oil-in-water (a) and water-in-oil (b) emulsions with IL-based surfactants.

After creating microemulsions, IL-based surfactants prevent droplet coalescence by forming protective layers with electrostatic and steric stabilization capabilities. In this regard, by covering the surface of the droplets with IL-based surfactants, their surface becomes electrically charged, so the positive or negative charges of the droplets separate them electrostatically and keep them stable. Also, the bulky and large IL-based surfactant head group sterically prevents the coalescence of droplets effectively. It is important to note that IL-based surfactants, due to their strong cohesive forces, could form stable microemulsions without co-surfactant, which are volatile and pose environmental hazards [8].

Research has shown that micellar flooding is an important CEOR process for depleted reservoirs. Microemulsions in reservoirs generally contain a transparent mixture of crude oil, an aqueous phase, and surfactants (sometimes with co-surfactants) in which oil, water, or both are dispersed [7]. In this regard, the study of various imidazolium IL-based surfactant microemulsions confirms their remarkable suitability to help to solubilize crude oils and create stable microemulsions to increase displacement and sweep efficiency [8]. Similarly, [C_8mim][BF_4], [C_{10}mim][BF_4], and [C_{12}mim][BF_4] can form Winsor III type microemulsion [45]. This feature is so appropriate because core flooding experiments confirm that the aqueous and oil phases coexisted in the intermediate phase of Winsor III microemulsions have the ultralow IFT value (less than 10^{-3} mN.m^{-1}), which leads to noteworthy CEOR in micellar flooding [7].

5.1 Effect of ionic liquid-based surfactants structure

The molecular structure of IL-based surfactants an influences the morphology and droplet size of the microemulsions. In terms of the alkyl chain length, interface disruption happens more easily with longer chain IL-based surfactants, resulting in producing small droplets. For instance, emulsion experiments with [$C_n$mim][BF_4] (n = 8, 10, and 12), confirmed that longer alkyl chains lead to more stable microemulsions [45]. In another work, it has been confirmed that [C_{16}mim][Cl], as a long chain imidazolium IL-based surfactant, can produce highly stable microemulsions [46]. In the same manner, longer alkyl chains in the anionic portions, by improving the amphiphilic nature, contribute to microemulsion formation. Accordingly, by studying the formation of microemulsions with [C_2mim][AS] (n = 2, 4, 6, and 8), it was shown that longer alkyl chain anions create more stable microemulsions [47].

The head group of IL-based surfactants plays a major role in determining whether microemulsions are formed in a polar or non-polar phase. In this regard, investigations on three Catanionic IL-based surfactants show that [C_4mim][AOT] creates stable microemulsions in both the aqueous and oil phases; though, [CTA][AOT] and [BHD][AOT] with hydrophobic head groups can only produce vesicles in the aqueous phase and microemulsions in the oil phase [48].

By changing the polarity of the counter anions, the effectiveness of IL-based surfactants on microemulsion formation changes. The bulkiness counter anions with greater polarizability provide better and more stable microemulsions. In this regard, the results show that the imidazolium IL-based surfactants work in the order [C_4mim][NTf_2] > [C_4mim][PF_6] > [C_4mim][BF_4] [49]. The counter anions, depending on their nature, can also affect the phase in which the microemulsions are formed, i.e. in the polar or non-polar phase.

5.2 New types of ionic liquid-based surfactants

Micellar flooding also strongly depends on the type of IL-based surfactants. Predictably, the formation and stability of microemulsions have been significantly enhanced with the newly introduced types of IL-based surfactants. In this regard, the study of the interfacial activity of the [P666,14][Cl], indicated that it provided stable microemulsions with no co-surfactant even at high temperatures [50, 51]. In another work, using the novel Anionic IL-based surfactant of [HOPro][DBS] reduced the IFT to an ultra-low value of 3×10^{-3} mN.m^{-1}, resulting in a highly stable microemulsion [52]. The results of microemulsions with Catanionic IL-based surfactant of [C_{12}mim][$(i\text{-}C_8)_2PO_2$] confirm their excellent performance even at low concentrations [53].

Tripodal and Gemini IL-based surfactants as a type of Multicationic, create stable microemulsions of crude oil in aqueous solutions. In this respect the effect of two benzimidazolium Gemini IL-based surfactants [C_4benzim-C_n-benzimC_4][Br_2], n = 4 and 6 are shown in **Figure 12**. As it is possible to see, stable microemulsions were obtained by them after one day and one week [25]. Similar results were achieved for imidazolium Gemini IL-based surfactants, which also improved with alkyl chain length [54, 55].

In another study, the performance of imidazolium Tripodal IL-based surfactant [$(C_4im)_3$TA][Cl_3] was compared with three imidazolium Gemini IL-based surfactants of [C_4im-C_n-imC_4][Br_2] n = 2, 4 and 6. The results are shown in **Figure 13**. As it is possible to see, the Tripodal IL-based surfactant also presents a greater emulsification performance even at an order of magnitudes lower concentration, which is due to superior hydrophobicity that provides easier interface disrupting and, consequently, the easier formation of smaller droplets [24].

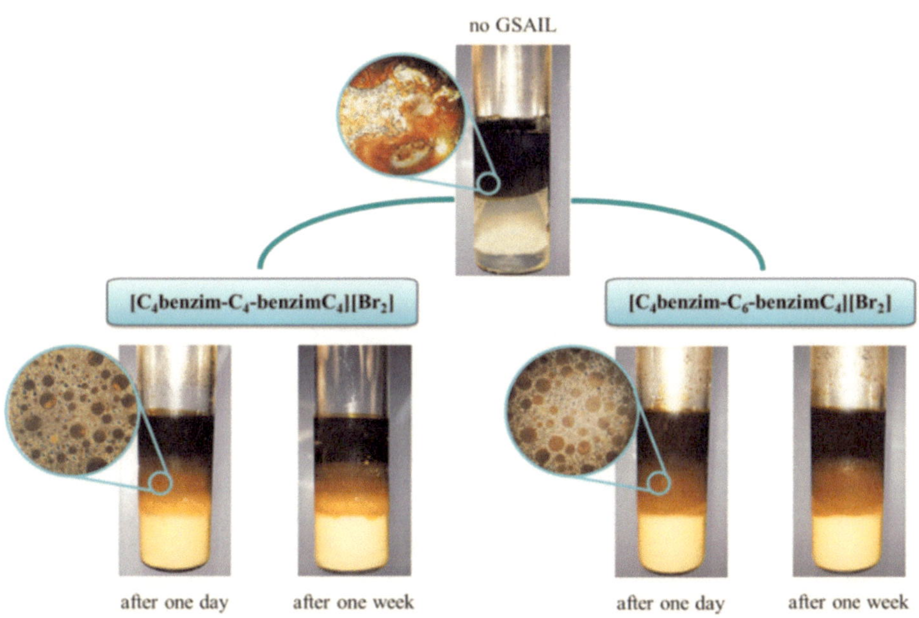

Figure 12.
The crude oil-water emulsions and the microscopic images (45 × magnification) with no Gemini IL-based surfactants and with 0.01 mol·dm^{-3} of them after one day and after one week [25].

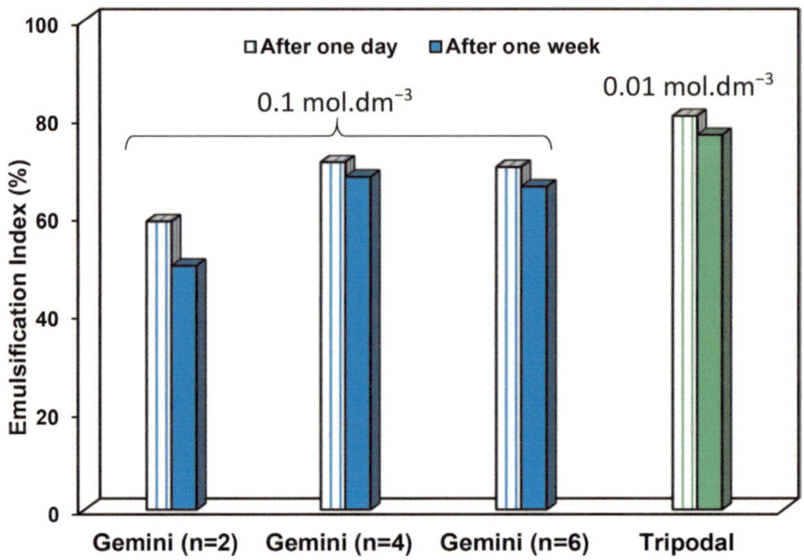

Figure 13.
The emulsification indices for 0.01 mol·dm^{-3} of the Tripodal and 0.1 mol·dm^{-3} of the Gemini IL-based surfactants after one day and one week [24].

6. Wettability alteration

Wettability, as determined by the interaction between rock surfaces and reservoir fluids (crude oil or brine), is recognized as one of the key parameters controlling the remaining original oil in place (OOIP). The wettability alteration in CEOR, which is related to the process of converting the reservoir rock from oil-wet to water-wet, remarkably increases the final oil recovery from the reservoirs. Likewise, wettability can be evaluated from the contact angle between the solid surfaces and the fluids as well as the three involved IFTs in the phase boundaries. The relationship between the contact angle and the involved three IFTs is determined using the well-known Young equation, which is presented for the solid-water-air (Eq. 1) and solid-water-oil (Eq. 2) systems as [56]:

$$\gamma_{s/a} - \gamma_{s/w} = \gamma_{w/a}\text{Cos}\theta \tag{1}$$

$$\gamma_{s/o} - \gamma_{s/w} = \gamma_{w/o}\text{Cos}\theta \tag{2}$$

In these Equations, θ is the contact angle and the various subscripts of s, o, w, and a, indicate the solid, oil, water, and air phases, respectively.

Solids with a contact angle less than 90° are hydrophilic surface (water-wet), solids with a contact angle of 90° have moderate behavior (intermediate-wet), and solids with a contact angle greater than 90° are hydrophobic (oil-wet). The contact angle for different wettability conditions of the solid-water-oil system is shown schematically in **Figure 14**. It is worth mentioning that for the oil droplets formed in the aqueous phase and in contact with the solid surface, the external contact angle with respect to the aqueous phase is considered.

In the wettability alteration, the adsorption of the IL-based surfactants on the surface of reservoirs makes the oil-wet rocks partially or completely water-wet.

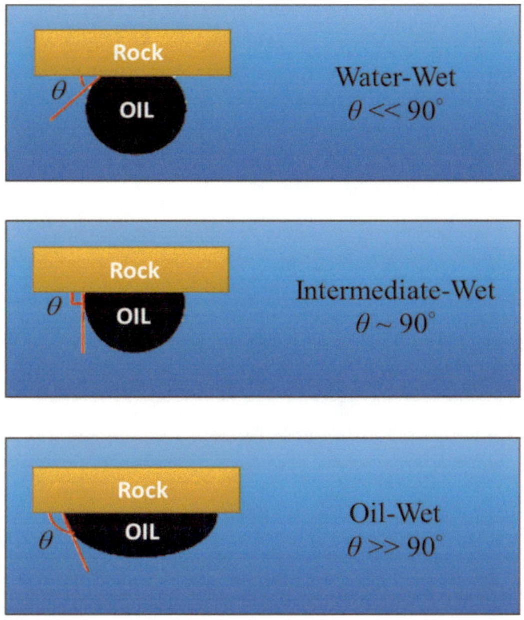

Figure 14.
Schematics of contact angle for different wettability conditions of the solid-water-oil system.

The mechanism by which IL-based surfactants alter the wettability of conventional rock pores is termed a cleaning mechanism, whereby IL-based surfactants desorb crude oil from the rock surface, so more oil can be extracted. The charge density, structure, and concentration of the IL-based surfactants as well as the surface charge of the rocks and the nature of the crude oil affect this procedure.

In carbonate reservoirs, negatively charged organic components of crude oil are adsorbed on the mineral surface by positively charged rock pores. When Cationic IL-based surfactants are added, ion-pair creation happens between the IL-based surfactants and the adsorbed anionic components of the crude oil (mainly carboxylate). This phenomenon repels crude oil from the rock surface by neutralizing the electrical charge of the adsorbed material (**Figure 15a**). Then, water imbibition takes place and the wettability of the rock becomes water-wet. In contrast, Anionic IL-based surfactants are unable to remove the crude oil carboxylates from the rock surface due to similar surface charges. As an alternative, as presented in **Figure 15b**, the hydrophobic interaction between the tail of the IL-based surfactants and the adsorbed crude oil on the surface of the reservoir rocks, such that the polar parts of the IL-based surfactants are directed toward the solution, this configuration hydrophilizes the rock surface. Though these configurations displace the crude oil from the rock surface, the interaction is not as efficient as the ion-pair interaction and only alterations the rock surface from oil-wet to intermediate-wet [57]. Generally, the Cationic IL-based surfactants are more effective than other types of wetting agents in carbonate reservoirs.

The proposed mechanism of wettability modification by the IL-based surfactants on the carbonate surfaces could be used to clarify the wettability alteration of the oil-wet sandstone rock (with negatively charged), albeit in reverse. So, Anionic IL-based surfactants are more effectual for sandstone reservoirs.

Regarding the concentration effect, a study on the wettability alteration of crude oil-water–quartz system in the presence of imidazolium IL-based surfactants shows

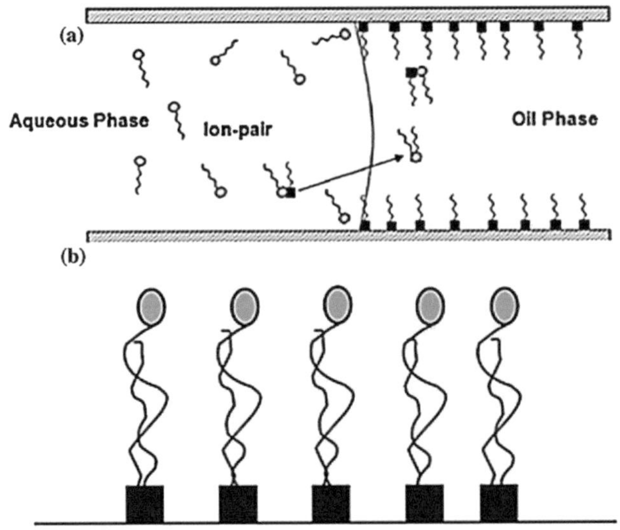

Figure 15.
Wettability alteration by cationic (a) and anionic IL-based surfactants (b) in carbonate reservoirs [57].

that with increasing IL-based surfactant concentration, the contact angle decreases rapidly in direction of changing wettability from oil-wet to water-wet [58].

6.1 Effect of ionic liquid-based surfactants structure

The change in the structure of IL-based surfactants meaningfully affects the wettability alteration. In this regard, the length of the alkyl chain appears to have a greater impact. Based on this, investigations on the effect of different imidazolium IL-based surfactants of [$C_n$mim][X] with n = 4, 6, and 8, and X = $[Cl]^-$, $[Br]^-$ and $[HSO_4]^-$, on the wettability of quartz surface in contact with model oil (including pentane, hexane, heptane, octane, decane, and dodecane) were carried out. The results revealed that IL-based surfactants with longer hydrocarbon chains switch wettability from oil-wet to water-wet better than shorter ones and provide lower contact angles [59]. Furthermore, the change of wettability in the presence of [$C_n$mim][BF_4] with n = 8, 10, and 12, indicated that as the chain length of IL-based surfactant increases, the contact angle decreases and leads to more favorable oil extraction [45].

The ability of the IL-based surfactant head groups to change carbonate surface wettability has also been investigated by measuring the contact angle of the oil droplets in the presence of two long alkyl chain IL-based surfactants of [C_{18}Py][Cl] and [C_{18}mim][Cl]. The results show that imidazolium IL-based surfactants are better than pyridinium ones in changing wettability to water-wet [30].

6.2 New types of ionic liquid-based surfactants

It is reasonably expected that new types of IL-based surfactants will have a fundamental effect on the wettability alteration. But unfortunately, little work has been done to investigate the performance of these new types in this field. In general, IL-based surfactants with different charges have different attractions and repulsions with the rock surfaces. Therefore, Cationic IL-based surfactants are preferred for use in carbonate reservoirs, while Anionic IL-based surfactants are suitable for sandstone

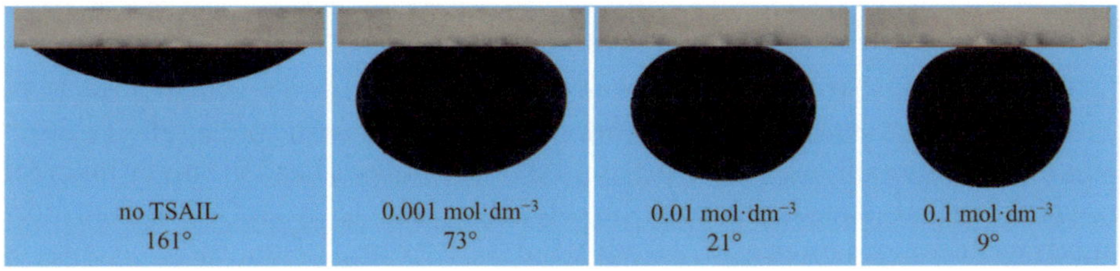

Figure 16.
The shapes of the crude oil drops on the quartz surface surrounded with different aqueous solutions of $[(C_4benzim)_3TA][Cl_3]$ [43].

reservoirs. So, appropriate IL-based surfactants should be selected based on the characterization of the reservoir's rock structure.

Investigations indicated better performance of Catanionic as well as Multicationic IL-based surfactants than other types of IL-based surfactants. In this regard, Gemini and Tripodal IL-based surfactants, as a type of Multicationic, significantly reduce the contact angle of crude oil in quarts surfaces so that even complete oil displacement with long chain Gemini IL-based surfactant $[C_{12}im\text{-}C_4\text{-}imC_{12}][Br_2]$ has been observed [38, 60]. For a better explanation, the change of contact angles of crude oil droplets on the quartz surface surrounded by different concentrations of $[(C_4benzim)_3TA][Cl_3]$ as a Tripodal IL-based surfactant is presented in **Figure 16**.

As another comparison, the performance of the benzimidazolium and imidazolium Tripodal IL-based surfactants of $[(C_4benzim)_3TA][Cl_3]$ and $[(C_4im)_3TA][Cl_3]$ as well as benzimidazolium Gemini IL-based surfactants $[C_4benzim\text{-}C_4\text{-}benzimC_4][Br_2]$ and $[C_4benzim\text{-}C_6\text{-}benzimC_4][Br_2]$ are shown in **Figure 17**. The results demonstrated that the benzimidazolium IL-based surfactants (specifically Tripodal) can greatly change the surface wettability, which confirms the higher hydrophobicity of the benzimidazolium rings [43].

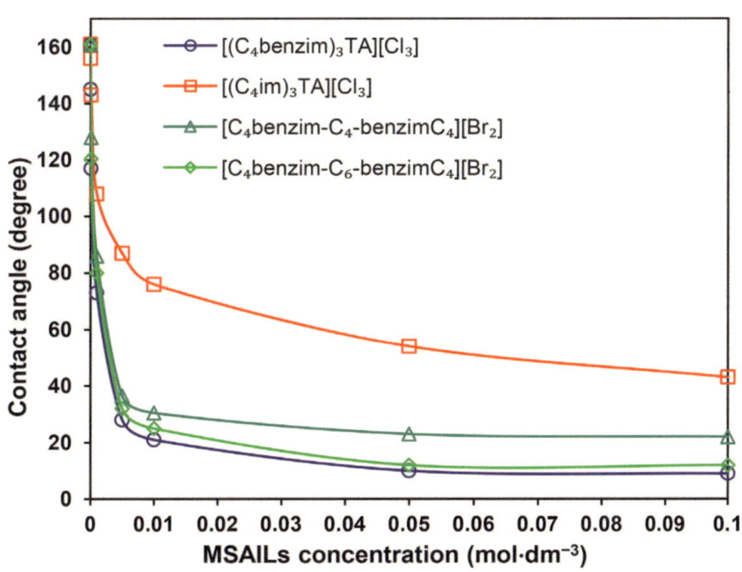

Figure 17.
The contact angles of crude oil drops on the quartz surface surrounded with multicationic IL-based surfactants solutions [43].

7. Ionic liquid-based surfactants versus conventional surfactants

In almost all cases, IL-based surfactants are more efficient in reducing the IFT of the crude oil-water system than conventional surfactants with similar alkyl chain lengths and counter anions. This is attributed to the bulkier head group or heteroatom ring, which gives IL-based surfactants greater hydrophobicity and easier transport to the interface; i.e. more interfacial activity. For better comparison, the molecules of a typical IL-based surfactant as well as a typical cationic surfactant are illustrated in **Figure 18**.

From a practical point of view, this is useful in CEOR processes as it allows the use of low amounts of IL-based surfactants. Correspondingly, the efficiency of IL-based surfactants can be enhanced by choosing suitable alkyl chains or anions as well as larger head group, preferably nitrogen-containing heterocyclic rings with delocalized charges. For comparison, the effects of 11 IL-based surfactants from three different families (lactam, ammonium, and hydroxylammonium) show that they could reduce the IFT better than Tergitol 15-S-5, Tergitol 15-S-7, and Tergitol 15-S-9 as conventional non-ionic surfactants [22, 56]. Also, the maximum reduction of crude oil-water IFT with [C_{16}mim][Br] and [C_{12}mim][Cl] was about 72%, which is higher than CTAB and DTMAC surfactants with similar structures but different head groups, whose reduce the IFT to about 68% [23].

It has to mention that some conventional surfactants may have a greater impact on IFT reduction even to very low levels. The performance of Cationic IL-based surfactants in IFT reduction can reach 10^{-2} mN.m^{-1} orders; however, some conventional surfactants can reduce IFT to about 10^{-3} (sometimes to 10^{-4}) mN.m^{-1}. To solve this problem, ultra-low IFT values can be achieved by developing the structure of IL-based surfactants as well as using Zwitterionic, Catanionic, Multicationic, or Anionic IL-based surfactants. For instance, the Catanionic IL-based surfactant of [P666,14][(i-C_8)$_2$$PO_2$] has shown a greater IFT reduction for the water-model oil system than conventional surfactants [61]. Moreover, the Anionic IL-based

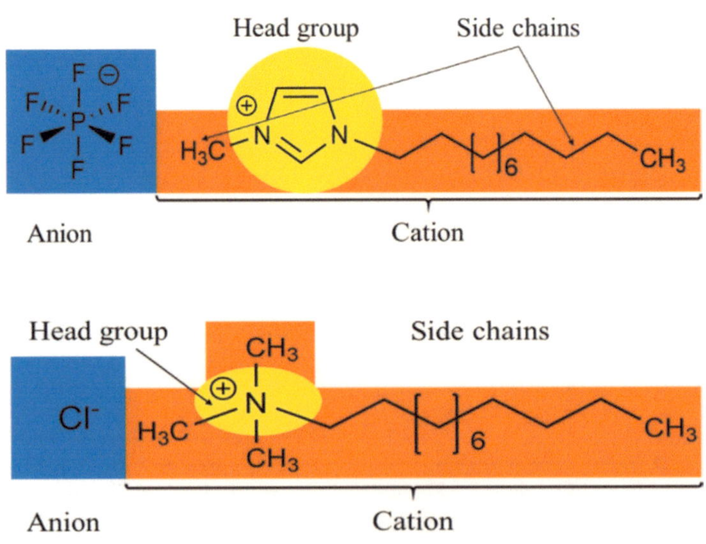

Figure 18.
Molecular structures of an imidazolium IL-based surfactant, [C_{12}mim][PF_6], and a conventional cationic surfactant, [DTMAC].

surfactant of [C_4mim][DS] effectively reduced the IFT of the water-benzene system to 0.9 mN.m^{-1}; whereas, by using SDBS and SDS as two conventional Anionic surfactants, this value was, respectively, about 5 and 8.5 mN.m^{-1}, respectively [62, 63]. In addition, the long chain Multicationic IL-based surfactants of [C_{12}im-C_4-imC_{12}][Br_2] can reduce the IFT of the crude oil-water system from 32 to about 10^{-3} mN.m^{-1} [38]. According to these results, newly introduced types of IL-based surfactants can be the next generation of surfactants with distinct properties, particularly in achieving ultra-low IFT values.

Regarding micellar flooding, conventional surfactants are difficult to use in micellar flooding CEOR because they not only tolerate high temperature and salinity conditions but also form Winsor II type microemulsions even at low salinity. Notably, investigations have confirmed that the IL-based surfactants can maintain their emulsifier performance under harsh reservoir conditions and can also form Winsor III type microemulsions, which is an interesting phenomenon in CEOR processes. As another advantage, IL-based surfactants have lower CMC compared to conventional analog surfactants, which is a significant point in micellar flooding. This is attributed to the bulkiness of heterocyclic head groups that give IL-based surfactants more hydrophobicity and greater aggregation tendency. These properties are economically significant. For example, imidazolium IL-based surfactants exhibit lower CMCs than the alkyltrimethylammonium bromides (C_nTAB) with the same chain length and anionic portion, but CMC of this type of IL-based surfactants are comparable to the Anionic surfactants, such as sodium alkyl sulfates (SAS) [13]. Likewise, the CMCs of [C_{16}mim][Br] and [C_{12}mim][Cl] have been 81 and 63% of those of DTMAC and CTAB, respectively [23]. Consistently, the Catanionic IL-based surfactants exhibit a lower CMC compared to conventional surfactants. Another important feature of IL-based surfactants is that the creation of stable microemulsions based on the use of conventional surfactants often requires co-surfactants, the most common of which are volatile and environmentally hazardous medium chain alcohols. However, IL-based surfactants, due to their strong cohesive forces, can form stable microemulsions without the need for co-surfactant requirements.

Moreover, IL-based surfactants are more efficient in wettability alteration from oil-wet to water-wet. In this regard, comparing the performance of imidazolium IL-based surfactant [C_{12}mim][Cl] with some conventional surfactants of AN-120, NX-610, NX-1510, NX-2760, and TR-880, by contact angle measurements, showing that IL-based surfactants are superior in changing the wettability from oil-wet to water-wet [64]. Therefore, it can be concluded that apart from surfactant flooding and micellar flooding, wettability alteration also play an important role in improving oil recovery with IL-based surfactants in CEOR processes.

The presented results clearly indicate that the IL-based surfactant is more effective than the conventional surfactants in the recovery of residual oil under reservoir conditions. Based on these results, IL-based surfactants can be considered effective alternative surfactants for CEOR processes.

8. Conclusions and future perspective

The present chapter provides an overview of the properties of IL-based surfactants and highlights their effectiveness in the oil industry. The results clearly show that the IL-based surfactants can be considered noteworthy candidates in the CEOR processes due to their high level performances in surfactant flooding and micellar flooding as well as wettability alteration that enhances crude oil production.

Surfactant flooding indicates that the IL-based surfactants have high interfacial activity and could reduce the IFT to a low value even under reservoir conditions. In view of the IL-based surfactant structure, increasing the length of the alkyl chain of cation and/or anion and the larger heterocyclic head group with π-electron ring, as well as bulkiness counter anion with higher polarizability, provides the more interfacial activity. Moreover, the new types of Catanionic, Multicationic, and Zwitterionic usually show higher interface activity compared to single analogs Cationic IL-based surfactants.

Furthermore, the study of IL-based surfactant microemulsions confirms their remarkable suitability to help solubilize crude oils and create stable microemulsions to enhance crude oil displacement and mobility. The molecular structure of IL-based surfactants influences the morphology and droplet size of the microemulsions. With longer chain IL-based surfactants, interface disruption occurs more easily, resulting in smaller droplets as well as more stable microemulsions. However, the head group plays the main role in determining the phase polarity of microemulsions formation, i.e. in the polar or non-polar phase. By increasing the polarity of the counter anion, the effectiveness of IL-based surfactants on microemulsion formation increases. The formation and stability of microemulsions are significantly enhanced by new types of IL-based surfactants, especially with Catanionic and Multicationic IL based-surfactants.

In the wettability alteration, adsorption of IL-based surfactants on the reservoir surfaces converts oil-wet rocks to water-wet. In general, Cationic IL-based surfactants are more efficient in carbonate reservoirs, while Anionic IL-based surfactants are better in sandstone reservoirs. The structure and concentration of IL-based surfactants affect this procedure, so IL-based surfactants with longer hydrocarbon chains and bulkier sizes have greater performance. As expected, new types of IL-based surfactants show better performance in changing wettability.

Comparison of IL-based surfactants with conventional surfactants demonstrates that due to larger cations or heteroatom rings, most of the IL-based surfactants show better interfacial activity in all three regions of surfactant flooding, micellar flooding, and wettability alteration than conventional surfactants with similar structures. Indeed, in all investigated cases, Zwitterionic, Catanionic, Multicationic or Anionic IL-based surfactants are more efficient.

The content of this chapter indicates that the advantages of IL-based surfactants make them a suitable candidates for use in the oil industry. In consequence, it is reasonably anticipated that IL-based surfactants can solve various problems in the oil industry. However, before large scale application, additional data such as crude oil nature, rock type, the interaction of IL-based surfactants with the surface of reservoir rocks, and solubility of IL-based surfactants in brine must be generated to provide real insight into their performance and economics. Thus, the screening of the IL-based surfactants is a necessary step before using them for the CEOR process to correctly predict the behavior of the IL-based surfactants.

Acknowledgements

The authors wish to thank the Bu-Ali Sina University authorities for supporting this work.

Conflict of interest

The authors declare no conflict of interest.

List of abbreviations and nomenclature

CEOR	Chemical enhanced oil recovery
CMC	Critical micelle concentration
EOR	Enhanced oil recovery
IFT	Interfacial tension
IL	Ionic liquid
OOIP	original oil in place
O/W	Oil-in-water
RTIL	Room temperature ionic liquid
SARA	Saturates, aromatics, resins and asphaltenes
Surfactant	Surface-active agent
W/O	Water-in-oil
θ	Contact angle
Cation part of IL-based surfactants	
[BHD]	Benzyl-n-hexadecyldimethylammonium
[BT]	Butyrolactam
[$C_n$mim]	1-Alkyl-3-methylimidazolium
[C_2mim]	1-Ethyl-3-methylimidazolium
[C_4mim]	1-Butyl-3-methylimidazolium
[C_6mim]	1-Hexyl-3-methylimidazolium
[C_8mim]	1-Octyl-3-methylimidazolium
[C_{10}mim]	1-Decyl-3-methylimidazolium
[C_{12}mim]	1-Dodecyl-3-methylimidazolium
[C_{16}mim]	1-Hexadecyl-3-methylimidazolium
[C_{18}mim]	1-Octadecyl-3-methylimidazolium
[C_4im-C_n-imC_4]	3,3′-(Alkane-1,4-diyl)bis(1-butyl-1*H*-imidazol-3-ium)
[C_8im-C_4-imC_8]	3,3′-(Butane-1,4-diyl)bis(1-octyl-1*H*-imidazol-3-ium)
[C_{12}im-C_4-imC_{12}]	3,3′-(Butane-1,4-diyl)bis(1-dodecyl-1*H*-imidazol-3-ium)
[C_4benzim-C_n-benzimC_4]	3,3′-(Alkane-1,4-diyl)bis(1-butyl-1*H*-benzo[*d*]imidazol-3-ium)
[C_4benzim-C_4-benzimC_4]	3,3′-(Butane-1,4-diyl)bis(1-butyl-1*H*-benzo[*d*]imidazol-3-ium)
[C_4benzim-C_4-benzimC_4]	3,3′-(Hexane-1,6-diyl)bis(1-butyl-1*H*-benzo[*d*]imidazol-3-ium)
[(C_4im)$_3$TA]	3,3′,3″-(1,3,5-Triazine-2,4,6-triyl)tris(1-butyl-1*H*-imidazol-3-ium)
[(C_4benzim)$_3$TA]	3,3′,3″-(1,3,5-Triazine-2,4,6-triyl)tris(1-butyl-1*H*-benzo[*d*]imidazol-3-ium)
[C_{16}mpyr]	1-Hexadecyl-1-methylpyrrolidinium
[C_8py]	1-Octylpyridinium
[C_{12}py]	1-Dodecylpyridinium

[C_{18}py]	1-Octadecylpyridinium
[Et_2NH_2]	Diethylammonium
[Et_3NH]	Triethylammonium
[CP]	Caprolactam
[CTA]	Cetyltrimethylammonium
[HOPro]	Prolinolium
[PA]	Propylammonium
[P144,4]	Tributyl-methylphosphonium
[P666,14]	Trihexyl-tetradecylphosphonium
[3-HPA]	3-Hydroxypropylammonium
Anion part of IL-based surfactants	
[Ac]	Acetate
[AOT]	1,4-Bis(2-ethylhexyl) sulfosuccinate
[AS]	Alkyl sulfate
[BF_4]	Tetrafluoroborate
[CH_3COO]	Acetate
[$C_6H_{13}COO$]	Hexanoate
[DBS]	Dodecylbenzenesulfonate
[DS]	Dodecylsulfonate
[HCOO]	Formate
[HSO_4]	Hydrogen sulfate
[H_2PO_4]	Dihydrogen phosphate
[$(i\text{-}C_8)_2PO_2$]	Bis(2,4,4-trimethylpentyl) phosphinate
[NTf_2]	Bis(trifluoromethylsulfonyl)imide
[OTF]	Trifluoromethanesulfonate,
[PF_6]	Hexafluorophosphate
[SCN]	Thiocyanate
[TA]	Trifluoroacetate,
IL-based surfactants	
[C_{16}IPS]	3-(1-hexadecyl-3-imidazolio) propanesulfonate
[C_{16}IPS-Nsa]	3-(1-hexadecyl-3-imidazolio) propanesulfonateβ-naphthalene sulfonate
Conventional surfactants	
AN-120	Dodecyl benzene sulfonic acid + Nonylphenolethoxylate
CTAB	Cetyltrimethylammonium bromide
C_nTAB	Alkyltrimethylammonium bromides
DTMAC	Dodecyltrimethylammonium chloride
NX-610	Dodecyl dimethyl amine betaine
NX-1510	Dodecyl dimethyl amine betaine + Octadecyl amine ethoxylate
NX-2760	Dodecyl dimethyl amine betaine + Octadecyl amine ethoxylated betaine
SDBS	Sodium dodecylbenzenesulfonate
SDS	Sodium dodecyl sulfate
SAS	Sodium alkyl sulfates
Tergitol 15-S-5	Ethoxylated C11 ~ 14 secondary alcohol
Tergitol 15-S-7	Ethoxylated C11 ~ 14 secondary alcohol
Tergitol 15-S-9	Ethoxylated C11 ~ 14 secondary alcohol

Author details

Mona Kharazi* and Javad Saien
Faculty of Chemistry and Petroleum Science, Bu–Ali Sina University, Hamedan, Iran

*Address all correspondence to: kharazi.mona@yahoo.com; kharazy91@yahoo.com

References

[1] Liu ZX, Liang Y, Wang Q, Guo YJ, Gao M, Wang ZB, et al. Status and progress of worldwide EOR field applications. Journal of Petroleum Science and Engineering. 2020;**193**:107449. DOI: 10.1016/j.petrol.2020.107449

[2] Ahmadi Y, Mohammadi M, Sedighi M. Introduction to chemical enhanced oil recovery. Chemical Methods. 2022;**2022**:1-32. DOI: 10.1016/B978-0-12-821931-7.00002-X

[3] Kharazi M, Saien J, Asadabadi S. Review on amphiphilic ionic liquids as new surfactants: From fundamentals to applications. Topics in Current Chemistry. 2022;**380**:1-44. DOI: 10.1007/s41061-021-00362-6

[4] Saien J, Kharazi M. Properties and applications of surface active ionic liquids. In: Abidi N, editor. Properties and Applications of Ionic Liquids. 1st ed. New York: Nova Science Publishers; 2023. pp. 125-156

[5] Saien J, Kharazi M. Surface active ionic liquids in petroleum industry. In: Abidi N, editor. Properties and Applications of Ionic Liquids. 1st ed. New York: Nova Science Publishers; 2023. pp. 157-180

[6] Bera A, Agarwal J, Shah M, Shah S, Vij RK. Recent advances in ionic liquids as alternative to surfactants/chemicals for application in upstream oil industry. Journal of Industrial and Engineering Chemistry. 2020;**82**:17-30. DOI: 10.1016/j.jiec.2019.10.033

[7] Nandwani SK, Malek NI, Chakraborty M, Gupta S. Insight into the application of surface-active ionic liquids in surfactant based enhanced oil recovery processes—A guide leading to research advances. Energy & Fuels. 2020;**34**:6544-6557. DOI: 10.1021/acs.energyfuels.0c00343

[8] Saien J, Kharazi M, Pino V, Pacheco-Fernández I. Trends offered by ionic liquid-based surfactants: Applications in stabilization, separation processes, and within the petroleum industry. Separation & Purification Reviews. 2023;**52**:164-192. DOI: 10.1080/15422119.2022.2052094

[9] Bera A. Applications of ionic liquids as green solvents in enhanced oil recovery. Green Sustainable Process for Chemical and Environmental Engineering and Science. 2023;**2023**:125-144. DOI: 10.1016/B978-0-323-95156-2.00016-7

[10] Johnson KE. What's an ionic liquid? Interface-Electrochemical Society. 2007;**16**:38-41. DOI: 10.1149/2.F04071IF

[11] Moghimi M, Kharazi M, Hashemi M, Noshirvani N. Investigating the effect of different physical and chemical parameters on the quality and stability of blackberry-enriched anthocyanin pigments. Food Research. 2023;**2023**:1-22 FR-2301-1861 (R2)

[12] Saien J, Hashemi S. Long chain imidazolium ionic liquid and magnetite nanoparticle interactions at the oil/water interface. Journal of Petroleum Science and Engineering. 2018;**160**:363-371. DOI: 10.1016/j.petrol.2017.10.057

[13] Dong B, Li N, Zheng L, Yu L, Inoue T. Surface adsorption and micelle formation of surface active ionic liquids in aqueous solution. Langmuir. 2007;**23**:4178-4182. DOI: 10.1021/la0633029

[14] Sedrpoushan A, Hosseini-Eshbala F, Mohanazadeh F, Heydari M. Tungstate

supported mesoporous silica SBA-15 with imidazolium framework as a hybrid nanocatalyst for selective oxidation of sulfides in the presence of hydrogen peroxide. Applied Organometallic Chemistry. 2018;**32**:e4004. DOI: 10.1002/aoc.4004

[15] Fabre E, Murshed SS. A review of the thermophysical properties and potential of ionic liquids for thermal applications. Journal of Materials Chemistry A. 2021; **9**:15861-15879. DOI: 10.1039/D1TA03656D

[16] Karunakaran M, Villalobos LF, Kumar M, Shevate R, Akhtar FH, Peinemann KV. Graphene oxide doped ionic liquid ultrathin composite membranes for efficient CO_2 capture. Journal of Materials Chemistry A. 2017;**5**: 649-656. DOI: 10.1039/C6TA08858A

[17] Pino V, German-Hernandez M, Martin-Perez A, Anderson JL. Ionic liquid-based surfactants in separation science. Separation Science and Technology. 2012;**47**:264-276. DOI: 10.1080/01496395.2011.620589

[18] Saien J, Kharazi M, Asadabadi S. Adsorption behavior of short alkyl chain imidazolium ionic liquidsat n-butyl acetate+ water interface: Experiments and modeling. Iranian Journal of Chemical Engineering. 2015;**12**:59-74

[19] Sastry NV, Vaghela NM, Aswal VK. Effect of alkyl chain length and head group on surface active and aggregation behavior of ionic liquids in water. Fluid Phase Equilibria. 2012;**327**:22-29. DOI: 10.1016/j.fluid.2012.04.013

[20] Li H, Imai Y, Takiue T, Matsubara H, Aratono M. Effect and mixing of counter anions at the surface of aqueous solution of imidazolium-based ionic liquids. Colloids and Surfaces A: Physicochemical and Engineering Aspects. 2013;**427**:26-32. DOI: 10.1016/j.colsurfa.2013.02.062

[21] Weingärtner H. Understanding ionic liquids at the molecular level: Facts, problems, and controversies. Angewandte Chemie International Edition. 2008;**47**:654-670. DOI: 10.1002/anie.200604951

[22] Sakthivel S, Chhotaray PK, Velusamy S, Gardas RL, Sangwai JS. Synergistic effect of lactam, ammonium and hydroxyl ammonium based ionic liquids with and without NaCl on the surface phenomena of crude oil/water system. Fluid Phase Equilibria. 2015;**398**: 80-97. DOI: 10.1016/j.fluid.2015.04.011

[23] Saien J, Asadabadi S. Alkyl chain length, counter anion and temperature effects on the interfacial activity of imidazolium ionic liquids: Comparison with structurally related surfactants. Fluid Phase Equilibria. 2015;**386**:134-139. DOI: 10.1016/j.fluid.2014.12.002

[24] Kharazi M, Saien J, Torabi M, Zolfigol MA. Molecular design and applications of a nanostructure green Tripodal surface active ionic liquid in enhanced oil recovery: Interfacial tension reduction, wettability alteration, and emulsification. Petroleum Science. 2023:1-10. DOI: 10.1016/j.petsci.2023.07.010

[25] Kharazi M, Saien J, Shokri B, Torabi M, Zolfigol MA. A comparative study on the design and application of new nano benzimidazolium gemini ionic liquids for curing interfacial properties of the crude oil–water system. RSC Advances. 2023;**13**:15747-15761. DOI: 10.1039/D3RA01783D

[26] Saien J, Kharazi M, Yarie M, Zolfigol MA. A systematic investigation of a surfactant type nano gemini ionic liquid and simultaneous abnormal salts

effects on the crude oil/water interfacial tension. Industrial & Engineering Chemistry Research. 2019;**58**:3583-3594. DOI: 10.1021/acs.iecr.8b05553

[27] José-Alberto MH, Jorge A. Current knowledge and potential applications of ionic liquids in the petroleum industry. In ionic liquids: Applications and perspectives. INTECH. 2011;**2011**: 439-456

[28] Hussain SS, Kamal MS. Effect of large spacer on surface activity, thermal, and rheological properties of novel amido-amine cationic gemini surfactants. Journal of Molecular Liquids. 2017;**242**:1131-1137. DOI: 10.1016/j.molliq.2017.07.128

[29] Levitt DB, Jackson AC, Heinson C, Britton LN, Malik T, Dwarakanath V, et al. Identification and evaluation of high-performance EOR surfactants. SPE Reservoir Evaluation & Engineering. 2009;**12**:243-253. DOI: 10.2118/100089-PA

[30] Manshad AK, Rezaei M, Moradi S, Nowrouzi I, Mohammadi AH. Wettability alteration and interfacial tension (IFT) reduction in enhanced oil recovery (EOR) process by ionic liquid flooding. Journal of Molecular Liquids. 2017;**248**:153-162. DOI: 10.1016/j.molliq.2017.10.009

[31] Anderson WG. Wettability literature survey-part 1: Rock/oil/brine interactions and the effects of core handling on wettability. Journal of Petroleum Technology. 1986;**38**:125-144. DOI: 10.2118/13932-PA

[32] Khandoozi S, Sharifi A, Riazi M. Enhanced oil recovery using surfactants. Chemical. Methods. 2022;**2022**:95-139. DOI: 10.1016/B978-0-12-821931-7.00007-9

[33] Drelich J, Fang C, White C. Measurement of interfacial tension in fluid-fluid systems. Encyclopedia of Surface and Colloid Science. 2002;**3**: 3158-3163

[34] Painter P, Williams P, Lupinsky A. Recovery of bitumen from Utah tar sands using ionic liquids. Energy & Fuels. 2010;**24**:5081-5088. DOI: 10.1021/ef100765u

[35] Zeinolabedini Hezave A, Dorostkar S, Ayatollahi S, Nabipour M, Hemmateenejad B. Effect of different families (imidazolium and pyridinium) of ionic liquids-based surfactants on interfacial tension of water/crude oil system. Fluid Phase Equilibria. 2013;**360**: 139-145. DOI: 10.1016/j.fluid.2013.09.025

[36] Saien J, Kharazi M, Asadabadi S. Adsorption behavior of long alkyl chain imidazolium ionic liquids at the n-butyl acetate+water interface. Journal of Molecular Liquids. 2015;**212**:58-62. DOI: 10.1016/j.molliq.2015.08.056

[37] Kharazi M, Saien J, Yarie M, Zolfigol MA. Different spacer homologs of gemini imidazolium ionic liquid surfactants at the interface of crude oil-water. Journal of Molecular Liquids. 2019;**296**:111748. DOI: 10.1016/j.molliq.2019.111748

[38] Kharazi M, Saien J, Yarie M, Zolfigol MA. The superior effects of a long chain gemini ionic liquid on the interfacial tension, emulsification and oil displacement of crude oil-water. Journal of Petroleum Science and Engineering. 2020;**195**:107543. DOI: 10.1016/j.petrol.2020.107543

[39] Sakthivel S, Velusamy S, Gardas RL, Sangwai JS. Use of aromatic ionic liquids in the reduction of surface phenomena of crude oil–water system and their synergism with brine. Industrial &

Engineering Chemistry Research. 2015; **54**:968-978. DOI: 10.1021/ie504331k

[40] Saien J, Kharazi M. A comparative study on the interface behavior of different counter anion long chain imidazolium ionic liquids. Journal of Molecular Liquids. 2016;**220**:136-141. DOI: 10.1016/j.molliq.2016.04.028

[41] Rodríguez-Escontrela I, Rodríguez-Palmeiro I, Rodríguez O, Arce A, Soto A. Characterization and phase behavior of the surfactant ionic liquid tributylmethylphosphonium dodecylsulfate for enhanced oil recovery. Fluid Phase Equilibria. 2016; **417**:87-95. DOI: 10.1016/j.fluid.2016.02.021

[42] Jia H, Lian P, Liang Y, Zhu Y, Huang P, Wu H, et al. Systematic investigation of the effects of zwitterionic surface-active ionic liquids on the interfacial tension of a water/crude oil system and their application to enhance crude oil recovery. Energy & Fuels. 2018;**32**:154-160. DOI: 10.1021/acs.energyfuels.7b02746

[43] Kharazi M, Saien J, Torabi M, Zolfigol MA. Green nano multicationic ionic liquid based surfactants for enhanced oil recovery: A comparative study on design and applications. Journal of Molecular Liquids. 2023;**383**:122090. DOI: 10.1016/j.molliq.2023.122090

[44] Danielsson I, Lindman B. The definition of microemulsion. Colloids and Surfaces. 1981;**3**:391-392. DOI: 10.1016/0166-6622(81)80064-9

[45] Pillai P, Kumar A, Mandal A. Mechanistic studies of enhanced oil recovery by imidazolium-based ionic liquids as novel surfactants. Journal of Industrial and Engineering Chemistry. 2018;**63**:262-274. DOI: 10.1016/j.jiec.2018.02.024

[46] Zech O, Thomaier S, Bauduin P, Rück T, Touraud D, Kunz W. Microemulsions with an ionic liquid surfactant and room temperature ionic liquids as polar pseudo-phase. The Journal of Physical Chemistry B. 2008; **113**:465-473. DOI: 10.1021/jp8061042

[47] Ghosh S, Banerjee C, Mandal S, Rao VG, Sarkar N. Effect of alkyl chain of room temperature ionic liquid (RTILs) on the phase behavior of [C2mim][C n SO4]/TX-100/cyclohexane microemulsions: Solvent and rotational relaxation study. The Journal of Physical Chemistry B. 2013; **117**:5886-5897. DOI: 10.1021/jp400013r

[48] Rao VG, Ghosh S, Ghatak C, Mandal S, Brahmachari U, Sarkar N. Designing a new strategy for the formation of IL-in-oil microemulsions. The Journal of Physical Chemistry B. 2012;**116**:2850-2855. DOI: 10.1021/jp2110488

[49] Rao VG, Mandal S, Ghosh S, Banerjee C, Sarkar N. Ionic liquid-in-oil microemulsions composed of double chain surface active ionic liquid as a surfactant: Temperature dependent solvent and rotational relaxation dynamics of coumarin-153 in [Py][TF2N]/[C4mim][AOT]/benzene microemulsions. The Journal of Physical Chemistry B. 2012;**116**:8210-8221. DOI: 10.1021/jp304668f

[50] Lago S, Francisco M, Arce A, Soto A. Enhanced oil recovery with the ionic liquid trihexyl (tetradecyl) phosphonium chloride: A phase equilibria study at 75 C. Energy & Fuels. 2013;**27**:5806-5810. DOI: 10.1021/ef401144z

[51] Lago S, Rodríguez H, Khoshkbarchi MK, Soto A, Arce A. Enhanced oil recovery using the ionic liquid trihexyl (tetradecyl)

phosphonium chloride: Phase behaviour and properties. RSC Advances. 2012;**2**: 9392-9397. DOI: 10.1039/C2RA21698A

[52] Fernández-Stefanuto V, Corchero R, Rodríguez-Escontrela I, Soto A, Tojo E. Ionic liquids derived from proline: Application as surfactants. ChemPhysChem. 2018;**19**:2885-2893. DOI: 10.1002/cphc.201800735

[53] Hejazifar M, Earle M, Seddon KR, Weber S, Zirbs R, Bica K. Ionic liquid-based microemulsions in catalysis. The Journal of Organic Chemistry. 2016;**81**: 12332-12339. DOI: 10.1021/acs.joc.6b02165

[54] Kharazi M, Saien J. Mechanism responsible altering in interfacial tension and emulsification of the crude oil-water system with nano gemini surface active ionic liquids, salts and pH. Journal of Petroleum Science and Engineering. 2022;**219**:111090. DOI: 10.1016/j.petrol.2022.111090

[55] Kharazi M, Saien J. Upgrading the properties of crude oil–water system for EOR with simultaneous effects of a homologous series of nano gemini surface active ionic liquids, electrolytes and pH. ACS Omega. 2022;**7**: 40042-40053. DOI: 10.1021/acsomega.2c04741

[56] Bera A, Ojha K, Mandal A, Kumar T. Interfacial tension and phase behavior of surfactant-brine–oil system. Colloids and Surfaces A: Physicochemical and Engineering Aspects. 2011;**383**:114-119. DOI: 10.1016/j.colsurfa.2011.03.035

[57] Gbadamosi AO, Junin R, Manan MA, Agi A, Yusuff AS. An overview of chemical enhanced oil recovery: Recent advances and prospects. International Nano Letters. 2019;**9**:171-202. DOI: 10.1007/s40089-019-0272-8

[58] Markiewicz M, Mrozik W, Rezwan K, Thöming J, Hupka J, Jungnickel C. Changes in zeta potential of imidazolium ionic liquids modified minerals—Implications for determining mechanism of adsorption. Chemosphere. 2013;**90**:706-712. DOI: 10.1016/j.chemosphere.2012.09.053

[59] Velusamy S, Sakthivel S, Sangwai JS. Effect of imidazolium-based ionic liquids on the interfacial tension of the alkane–water system and its influence on the wettability alteration of quartz under saline conditions through contact angle measurements. Industrial & Engineering Chemistry Research. 2017;**56**: 13521-13534. DOI: 10.1021/acs.iecr.7b02528

[60] Kharazi M, Saien J, Yarie M, Zolfigol MA. Promoting activity of gemini ionic liquids surfactant at the interface of crude oil-water. Petroleum Research. 2021;**31**:20-22. DOI: 10.22078/PR.2020.4130.2874

[61] Rodríguez-Escontrela I, Rodríguez-Palmeiro I, Rodríguez O, Arce A, Soto A. Phase behavior of the surfactant ionic liquid trihexyltetradecylphosphonium bis(2,4,4-trimethylpentyl)phosphinate with water and dodecane. Colloids and Surfaces A: Physicochemical and Engineering Aspects. 2015;**480**:50-59. DOI: 10.1016/j.colsurfa.2015.04.002

[62] Zhou H, Zhu Y, Peng T, Song Y, An J, Leng X, et al. Systematic study of the effects of novel halogen-free anionic surface active ionic liquid on interfacial tension of water/model oil system. Journal of Molecular Liquids. 2016;**223**: 516-520. DOI: 10.1016/j.molliq.2016.08.080

[63] Saien J, Eghtenaie A, Kharazi M. Synergistic performance of a gemini ionic liquid and sodium dodecyl sulfate surfactants at the crude oil–water

interface. Arabian Journal of Chemistry. 2023. In Press

[64] Nabipour M, Ayatollahi S, Keshavarz P. Application of different novel and newly designed commercial ionic liquids and surfactants for more oil recovery from an Iranian oil field. Journal of Molecular Liquids. 2017;**230**: 579-588. DOI: 10.1016/j.molliq.2017.01.062

Chapter 6

Effect of Microheterogeneous Environments on Reaction Rate: Theoretical Explanation through Different Models

Harasit Kumar Mandal

Abstract

Microheterogeneous systems (typically micelles and reverse micelle systems) refer to aggregate systems in which different structural shapes are formed by amphiphiles in water or other similar media. Amphiphilic molecules undergo a self-assembly process known as micellization through hydrophobic or H-bonding interactions. The Microenvironment consisting surfactants changes the physicochemical behavior of the system and can catalyze or inhibit reaction rates. In addition, thermodynamic parameters such as standard enthalpy of activation, standard entropy of activation, energy of activation etc. may vary in this environment. Thus, for last few decades, the enormous interest among the researchers in the study of the effect of microheterogeneous environments on reaction rate. It is also challenging how to fit experimental results with different models. Every model has its own significance and limitations.

Keywords: surfactants, microheterogeneous environments, determinations of micelle, reactions in microheterogeneous environments, different kinetic models

1. Introduction

For last few decades, there has been enormous interest among the researchers in the study of the effect of micro-heterogeneous environments on reaction rate. Amphiphile molecules generally form such environments due to their hydrophilic head and hydrophobic tail groups, they can able to interact with both polar and nonpolar solvents (/compounds) and easily form micelle or reverse micelle structures [1]. Reactants accommodated in surfactant-based organized assemblies, such as micelles, microemulsions, and vesicles, often achieve a greater degree of organization compared to their geometries in homogeneous continuous solution, can mimic reactions in biosystems, and have potential for energy storage [2]. Micelles can cause an acceleration (/catalyze) or inhibition of a given chemical reaction relative to the equivalent reaction in reaction medium due to the concentration effect in the micellar pseudophase and can change the reaction pathway [3–5]. Reverse micelles also play an important role to alter the reaction pathway as well as rate of reactions. Surfactants in nonpolar solvents can solubilize considerable amounts of water with the formation of inverted or reversed micelles. The size of the spherical reverse micellar cavity known as the water pool (w = [H_2O]/[Surfactant]) [6, 7] carries the most important structural feature of the reverse micellar system and controls the reaction rate. Coulombic, hydrophobic and charge transfer can form a variety of interactions between the above-mentioned organized media and incoming species. All of these interactions can play an important role in altering rates and reaction pathways by altering the micro-environment around the reacting species. Therefore, micro heterogeneous environments are very important topic to many researchers. Thermodynamic parameters for the micelle formation (i.e. the standard free energy of micelle formation,G^0_m; the standard enthalpy of micelle formation, H^0_m etc.) obtained from the temperature dependence of the CMC and the degree of counter ion binding have been reported in several works [8, 9]. Most kinetic data in kinetic studies are interpreted with one of the proposed micellar models such as pseudo-phase (PP) model [3, 4], ion-exchange PP model [10], Columbic model [11], enzyme kinetic type model [12], mass-action and multiple separation micellar models [13, 14]. The chapter deals with the effect of micro-heterogeneous environments on reaction rates and briefly discusses various aspects of different kinetic models for treating micellar catalysis or inhibition on the rate of reactions.

2. Surfactants

Surfactants are surface-active agents, active on surfaces. They are compounds that lower the surface tension (/interfacial tension) due to the adsorption at fluid surfaces and interfaces and change the property of the interface considerably. Some example of surface and interface are solid-vapor (surface), liquid-vapor (surface), solid–liquid (interface), solid–solid (interface), liquid–liquid (interface) etc. They are usually organic compounds that are amphiphilic in nature (thus sometimes called amphipathic structure) that means they contain both tails (lyophobic groups) and heads (lyophilic groups). Their heads are polar, or hydrophilic, and their tails hydrophobic (**Figure 1**). They are soluble in both organic solvent and water. When the solvent is water, the lyophobic and lyophilic parts are called hydrophobic and hydrophilic, respectively. Surfactant is one of the most frequently used chemical in the everyday life with a wide range of industrial application such as in detergents [15], pharmaceuticals [16], cosmetics [17], textiles [18], medicine [19], chemistry [20] etc.

2.1 Chemical structure of surfactants

The aliphatic, aromatic hydrophobic part of the surfactant contains 8 to 22 carbons in linear or branched chain form. The hydrophobic part is usually obtained from fats, oils, synthetic polymers or synthetic alcohols. The surfactant molecules also have a functional group called the hydrophilic group or polar group. The choice of polar group and comparative size of the hydrophobic and polar groups are important factors in determining the properties and surfactant physiochemical behavior in water. Depending on the type of charge on the head groups attached to their chains, surfactants are mainly divided into two categories, ionic and non-ionic, also zwitterionic (or amphoteric) and Gemini surfactants. A very brief description of the different types of surfactants is given below:

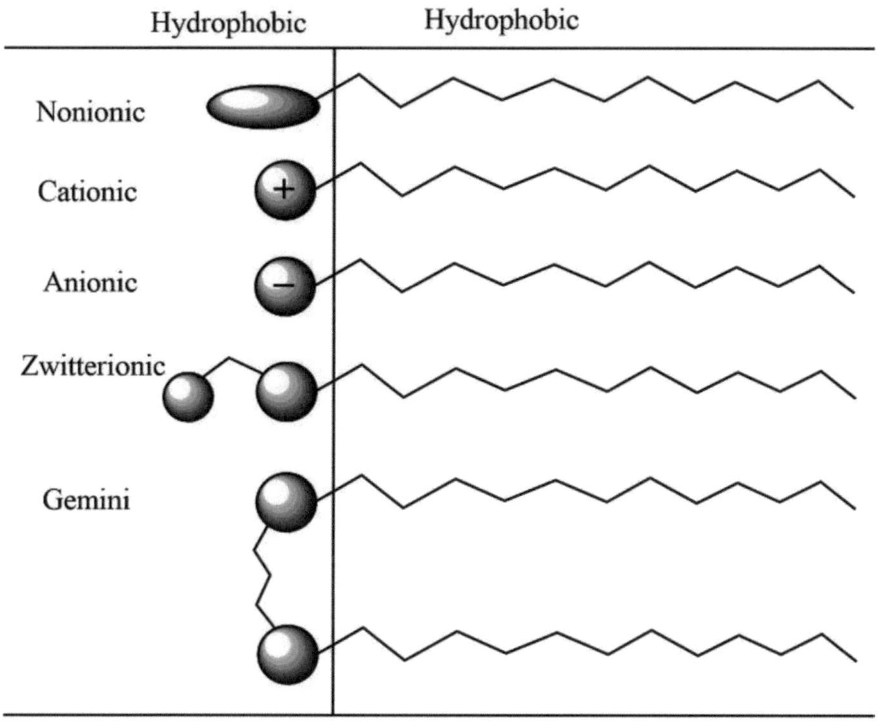

Figure 1.
Different types of surfactants: Their hydrophobic and hydrophilic parts.

2.1.1 Cationic surfactants

Cationic surfactants are substances whose head groups are positively charged. Thus, these are useful for adsorption on negatively charged surfaces. Some well-known cationic surfactants are long chain amines ($R\overset{+}{N}H_3X^-$), quaternary ammonium salts ($R\overset{+}{N}(CH_3)_3X^-$) and quaternary salts of polyethylene oxide amine derivatives ($R\overset{+}{N}(CH_3)\{(C_2H_4O)_xH\}_2Cl^-$). The most common cationic surfactants are shown schematically in **Figure 2**.

2.1.2 Anionic surfactants

The head groups (or functional groups) of anionic surfactants are negatively charged and neutralized by alkali metal cations. The head groups of anionic surfactants are based on carboxylates (e.g. soaps, $RCOO^-Na^+$), sulphates ($RSO_4^-Na^+$), phosphates ($RPO_4^{2-}2Na^+$) and benzene sulphonates ($RC_6H_4SO_3^-Na^+$). Soaps are the largest single type of anionic surfactants obtained by saponification of natural oils and fats. The use of anionic surfactants are very sensitive to hardness of water because some of the anionic surfactants (e.g. salts of fatty acids) are precipitated from the aqueous solution in presence of salts containing Ca^{2+} and Al^{3+} ions. **Figure 3** shows the most useful anionic surfactants.

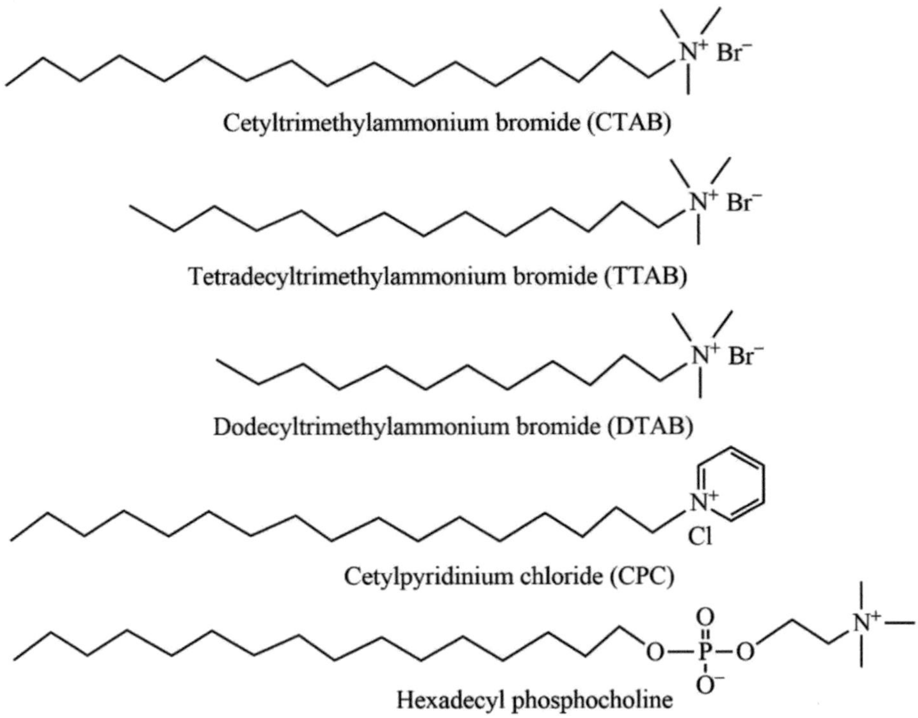

Figure 2.
Chemical structures of some common cationic surfactants.

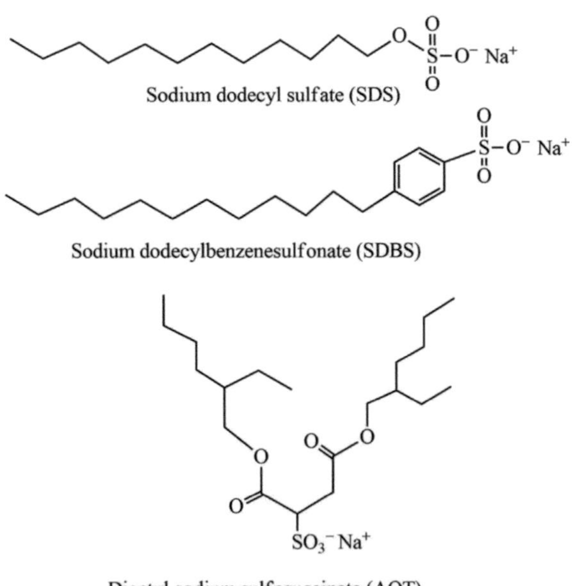

Figure 3.
Chemical structures of commonly used anionic surfactants.

2.1.3 Non-ionic surfactants

In these types of surfactants, they do not have any significant charge on their surface-active part. Therefore, there is no electrical interaction between the head groups. These surfactants are stable in presence of electrolysis. They usually generate less foam than the ionic surfactants. They are insensitive to water hardness. They are mainly used as dispersing agents for pigments in paints, foam control agents, also used in textile detergents, metal cleaning and shampoos. These surfactants have covalently bonded oxygen containing hydrophilic groups, which are bonded to hydrophobic parent structures (**Figure 4**). The common examples of non-ionic surfactants are polyoxypropylene glycols, polyoxyethylene mercaptants [RS $(C_2H_4O)_xH$], alkyl phenol ethoxylene [RC_6H_4 $(OC_2H_4)_xOH$]; alcohol ethoxylates [$R(OC_2H_4)_xOH$] etc.

2.1.4 Zwitterionic surfactants

Zwitterionic surfactant consists of two oppositely charged groups, both charges, positive and negative present on the surface-active part of the same molecule. The cationic part is based on primary, secondary, tertiary amines or quaternary ammonium cations (**Figure 5**). The anionic part can be more variable and include sulfonates, phosphates etc. Among them, the long chain amino acids ($RN^+H_2CH_2COO^-$) are the well-known examples of the zwitterionic surfactants (**Figure 5**). Their main advantage is that they are compatible with both anionic and cationic surfactants due to the presence of both positive and negative charges. Most of these surfactants are sensitive to pH. They are considered zwitterionic at pH 7. However, they show the

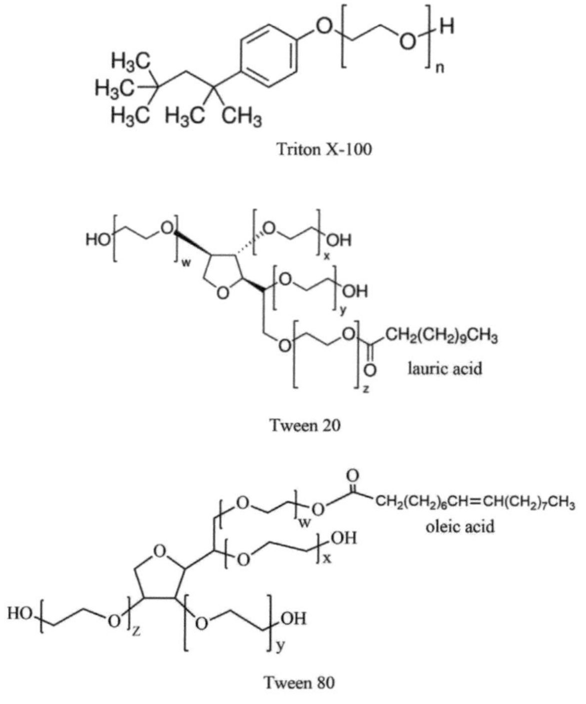

Figure 4.
Chemical structures of common non-ionic surfactants.

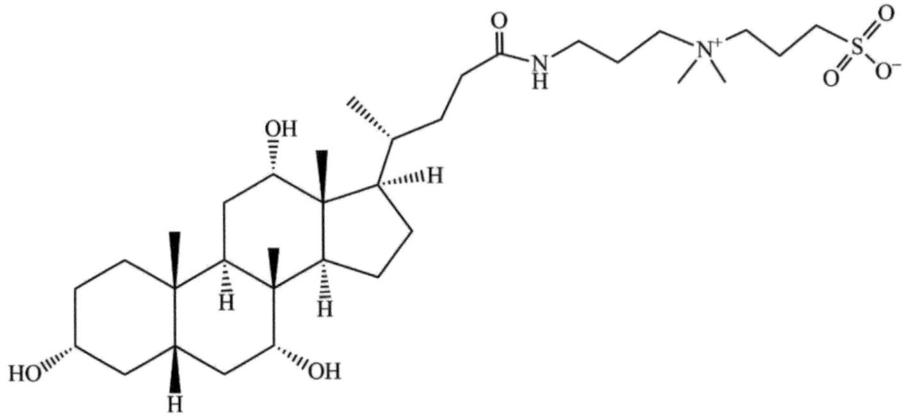

CHAPS

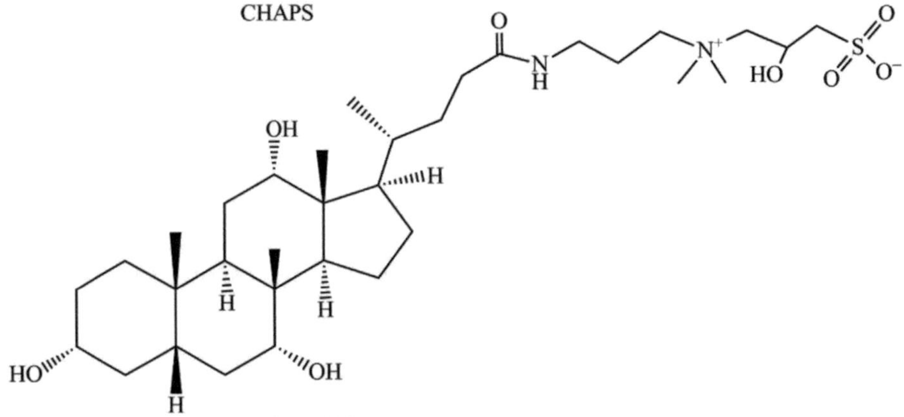

CHAPSO

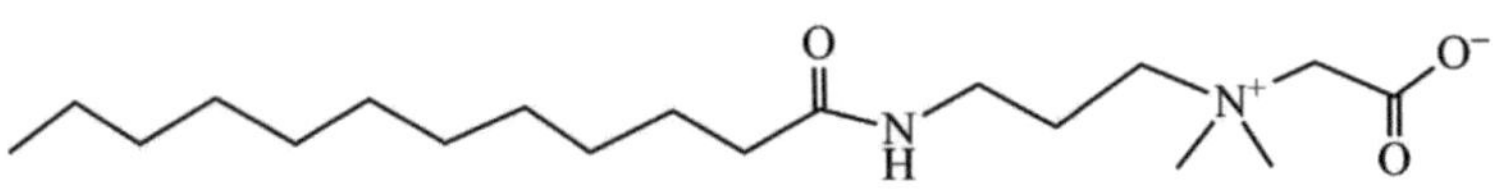

Cocamidopropyl betaine

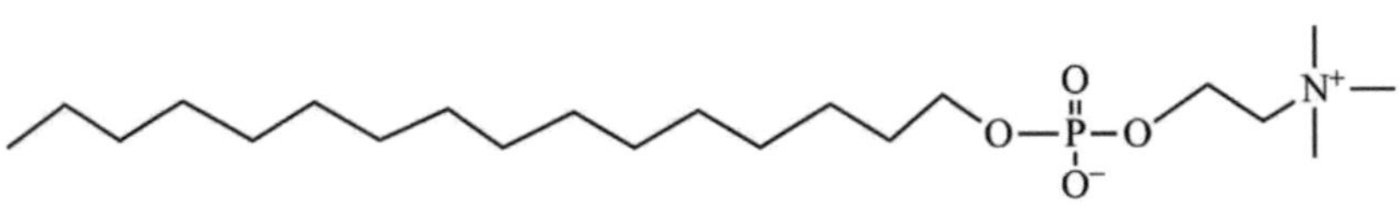

Hexadecyl phosphocholine

Figure 5.
Chemical structures of some zwitterionic surfactants.

properties of anionic surfactants at high pH whereas they behave as cationic surfactants at low pH.

2.1.5 Gemini surfactants

A surfactant contains one polar group. In recent times, there have been considerable research interest in certain dimeric surfactants, which contain two or three hydrophobic tails and two head groups linked together with a short spacer (**Figure 6**). They are referred to as Gemini surfactants. The properties of these surfactants are

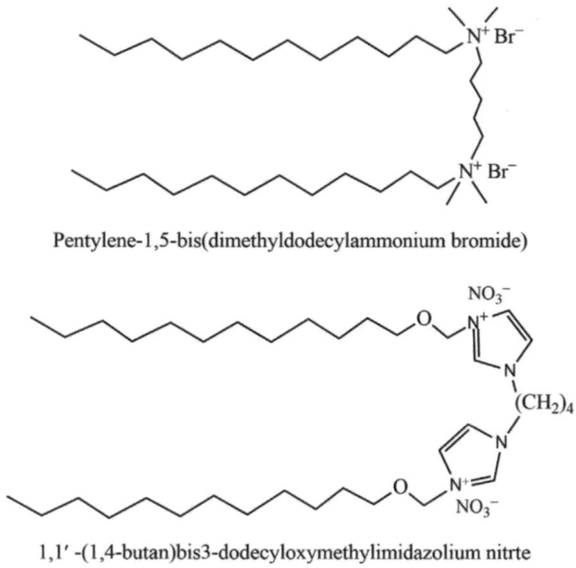

Figure 6.
Chemical structures of some Gemini surfactants.

highly dependent on the structure of these three parts of the molecule. The interfacial effects of these surfactants can be significantly greater than that of surfactants with single hydrophilic and hydrophobic groups [21]. Gemini surfactants can be ionic with negative, positive or both types of charge or can be non-ionic. These surfactants are less water soluble due to their long hydrophobic part and have an affinity for adsorption at the interface.

3. Surfactant aggregation (/self-assemble): Formation of micelle and reverse micelle

Surfactants have polar hydrophilic parts, called head groups, and non-polar hydrophobic parts, called tail groups. This amphipathic character of surfactants may produce ensembles viz., micelles, vesicles, fibers, discs, lamellae, bicontinuous phases, liquid crystals, and 3D networks [22–26]. In micelle, the outer core of it is made up of hydrophobic part while the polar head groups are positioned at the inner micelle-water interface (**Figure 7c**). The polarity of the medium is an important factor for surfactant association. Usually, micellization occur in aqueous solution. But in nonpolar medium (benzene, heptane, octane, cyclohexane etc.), formation of micelle is totally absent; rather a reverse orientation of surfactants occurs, where the tail groups are outside and head groups are inside the micelles, called ‘reverse micelles’ (**Figure 7d**). A trace amount of water helps easier and stable formation of reverse micelle. The other variations in shape in micelle clusters are lamellar, cylindrical, vesicle, liposome (**Figure 7**) etc.

Micelles are formed only above a minimum concentration that are characteristics of the system.

The lower limit of concentration of the surfactant at which micellization starts spontaneously is called as critical micelle concentration (CMC). The common names and CMCs of different surfactants are shown in **Table 1**. It is significant that micelle

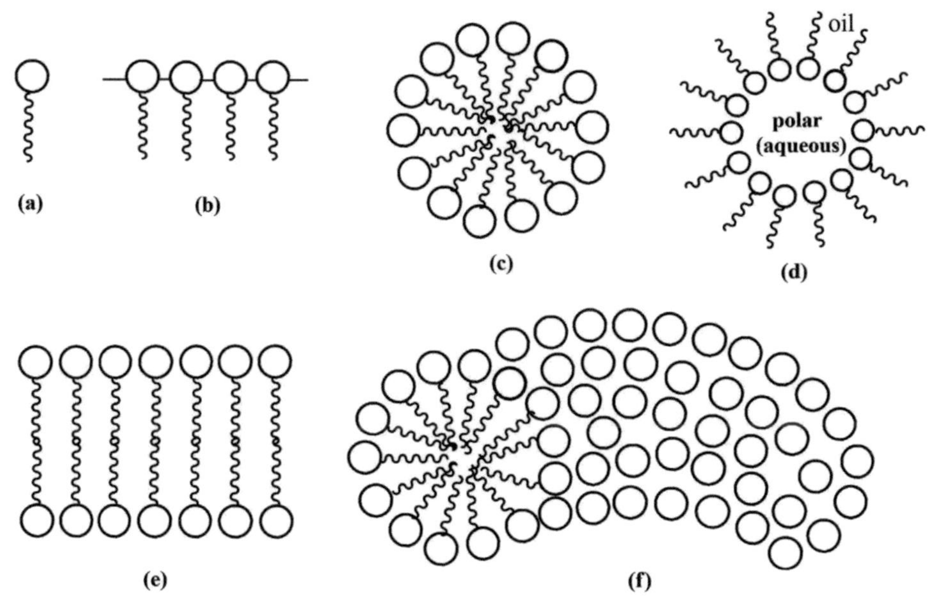

Figure 7.
(a) monomer, (b) monomer layer and different surfactant aggregations: (c) micelle (spherical shape) (d) reverse micelle (e) lamellar (f) micelle (cylindrical shape) (g) vesicle (h) liposome.

Surfactant type	Name and structure of surfactant	CMC (mol L^{-1})
Ionic surfactants	CTAB,	(0.98–1.04) × 10^{-3} [27, 28]
	TTAB,	(3.42–3.50) × 10^{-3} [4, 29, 30]
	DTAB,	(13.81–14.59) × 10^{-3} [3, 30, 31]
	CPC,	(0.9–1.03) × 10^{-3} [32, 33]
	SDS,	(6.81–8.6) × 10^{-3} [27, 34]
Non-ionic surfactants	AOT, sodium bis (ethylhexyl) sulfosuccinate	(2.66–2.68) × 10^{-3} [32, 35]
	Brij-35	(1.0–2.0) × 10^{-4} [36]
	Triton X 100	(0.25–0.27) × 10^{-3} [37, 38]
	Tween 20	(4.88–4.99) × 10^{-5} [39, 40]
	Tween 40	(2.26–3.33) × 10^{-5} [39, 40]
	Tween 60	(1.67–2.1) × 10^{-5} [39, 40]
	Tween 80	(1.5–1.89) × 10^{-5} [39, 41]

Surfactant type	Name and structure of surfactant	CMC (mol L^{-1})
Zwitterionic surfactants	CAPB,	(2.8–3.1) × 10^{-3} [42, 43]
	CHAPS, 3-[(3-Cholamidopropyl) dimethylammonio]-1-propanesulfonate hydrate	(4.0–5.65) × 10^{-3} [44]
Gemini surfactants	*Anionic gemini* Sodium2,3-didodecyl-1,2,3,4-butane-tetracarboxylate *Cationic gemini* 1,10-(alkane-1, s-diyl)bis(1-dodecyl pyrrolidinium)bromide (C_{12}-C_3-C_{12})PB	8.9 × 10^{-5} [45] (0.41–0.66) × 10^{-3} [46]

Table 1.
Essential surfactants and their CMC values.

formation becomes easier for surfactants with greater hydrophobicity (or increased chain length) in a homogeneous series. Other factors that can affect the formation of micelle are solvent polarity and type, temperature, pressure, presence of additives (e.g. salts) etc.

4. Structure of micelle

The hydrocarbon chain of the surfactant molecule can form the inner part of the micelle and its radius is about the range of a fully extended hydrophobic chain as shown in **Figure 8**. A stern layer covers the core and exists between the core surface and the hydrodynamic shear surface of ionic micelles. The stern layer contains the n fully ionized head groups and $(1-\alpha)$n counter ions, where α is the degree of ionization

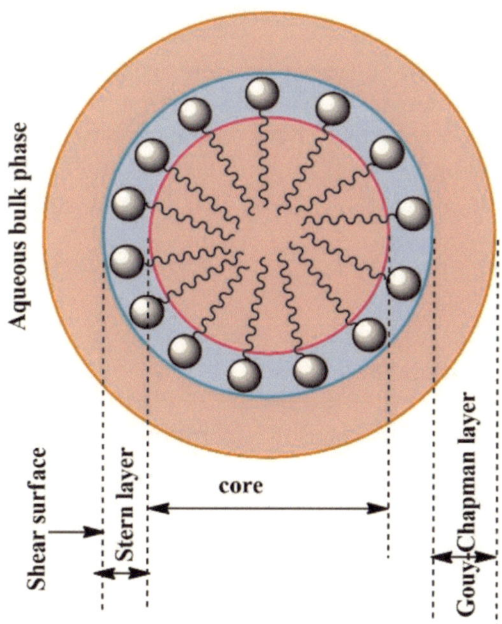

Figure 8.
A two dimensional schematic representation of a spherical ionic micelle.

and for most ionic micelles, it is 0.2 to 0.3; n is the aggregation number (the number of monomers required to form an aggregate is known as aggregation number) [47]. The remaining 'αn' counter ions neutralizing the micelle charge are distributed outside the shear surface in a Gouy-Chapman diffuse ionic atmosphere. The Gouy-Chapman layer is diffuse and surrounds the Stern layer i.e. the outer surface of the Stern layer.

The structure of micelles formed by non-ionic surfactants is almost identical to micelles formed by ionic surfactants except the counter ions. In ionic micelles, the counter ions are found in the outer regions, while in non-ionic micelles there are loops of hydrated polyethylene oxide chains. They are larger than ionic micelles and sometimes extend into ellipsoidal or rod-like structures. They contain a hydrophobic core, formed by the hydrocarbon chains of the surfactant, surrounded by a shell (palisade layer) made by the oxyethylene chains of the surfactant.

5. Structure of reverse micelle

In non-aqueous solution, the micelles are formed, known as reverse or inverted micelles. Reverse micelles are spherical molecular aggregates dispersed in an oil phase. The polar solvent within the droplet is covered by the polar head groups of surfactant (**Figure 7d**). For most of the systems, a co-surfactant (usually long chain alcohol) is added to increase the stability. The size of the spherical reverse micellar cavity corresponds to hydrodynamic radius (r_h) is reflected by the molar ratio of water to surfactant, which is known as water pool (w) [6].

$$w = \frac{[H_2O]}{[\text{Surfactant}]}$$

The temperature, nature of oil (non-polar solvent) and the dispersed phase volume fraction (ϕ) have slight effects on the size of reverse micellar cavity. The size of the water pool can be precisely controlled by w on the nanometer scale. The parameter, w is the key factor controlling the most important structural features of the reverse micellar system [48].

6. Factors affecting CMC

6.1 Nature of chain and counter ion

It is imperative that micelle formation becomes easier for surfactants having greater hydrophobicity (or increased chain length) in a homologous series. Increase in the number of carbon atoms in unbranched hydrocarbon chain leads to decrease in CMC, since

$$\log(\text{CMC}) = \text{A–B} \times \text{m}$$

Where, m is the number of carbon atoms in the hydrophobic chain [49]. Increased hydrophobicity provided by increasing chain length leads to an increase in micellar

Surfactant	Aggregation number
Cationic surfactants:	
$C_{10}H_{21}N^{+}(CH_3)_3Br^{-}$	36
$C_{12}H_{25}N^{+}(CH_3)_3Br^{-}$	50
$C_{14}H_{29}N^{+}(CH_3)_3Br^{-}$	75
Anionic surfactants:	
$C_{10}H_{21}SO_4^{-}Na^{+}$	50
$C_{12}H_{25}SO_4^{-}Na^{+}$	80

Table 2.
Effect of the hydrophobic chain length on aggregation number in water at 298 K.

size. The aggregation number increases with increase in hydrocarbon chain length of the surfactant molecules (**Table 2**). The CMC decreases with the increase in chain length.

The CMC also depends on the size of the hydrophilic group. As the size of the hydrophilic group gets larger, the repulsion between them increases. The change of counter ion to one of greater valence leads to a decrease in CMC and increase in aggregation number. An increase of CMC is noted with increase in hydrated radius.

6.2 Solvent polarity and type

The nature of the solvent can affect the CMC. The polarity of the medium favors surfactant association. Nonpolar medium offers environment similar to the surfactant tail so that their tendency of self-association is reduced. In a good nonpolar medium like cyclohexane, carbon tetrachloride, heptane, octane, decane etc., favors for reverse orientation of the surfactants instead of normal micelle.

6.3 Temperature

Temperature is an important factor that can affect the micellization phenomenon, although the effect of temperature on micellization may not be straightforward. The effect of temperature on the formation of micelles depends largely on how temperature affects the solubility and other behaviors of surfactants in solution. In general, the solubility of surfactants in water does not increase dramatically with temperature. Desolvation and change in solvent structure play an essential role in this regard. Typically, micelle formation is favored by an increase in temperature in the lower temperature range compared to the higher temperature range.

The formation of micelle may be prevented by a strong electrostatic repulsion of the desolvated head groups. The situation is complicated by the change in polarity of the medium at high temperature. At high temperature, the solution becomes cloudy due to phase separation resulting from desolvation of non-ionic surfactant polar head groups. The temperature at which the phenomenon begins is known as the cloud point. Their cloud points often refer the temperature stability of non-ionic surfactants. Aqueous solutions of non-ionic surfactant micelles exhibit thermoreversible phase separation phenomena on heating/cooling through a cloud point (Tc) [50]. Over the past few years, the cloud point phenomenon has been significantly applied in extraction and separation science. It has also been successfully used for the recovery of various species such as organic compounds, inorganic metal ions, biological analytes, etc. Watanabe et al. [51] illustrated the extraction of soluble metal [Zn(II)] chelates from aqueous solutions by cloud point extraction (CPE). When modulating the cloud point, it seems easier to reduce than to increase. Many common salts, such as NaCl, are highly effective cloud point suppressants. In contrast some salts, such as those containing I^{-1}, $[Fe(CN)_5NO]^{-2}$ and SCN^{-1} anions, can raise the cloud point.

The surfactants have characteristic temperature dependent phase behaviors, shown in **Figure 9**. The Kraft temperature is a point of phase change below which the surfactant (monomer) remains in **crystalline form**, even in aqueous solution (**Figure 9**, zone C). In zone A, only surfactant monomers occur in solution and in zone B, **monomers** remain in equilibrium with **micelles**. At the point P, all the three phases (monomer, crystal and micelle) co-exist, known as invariant point and the corresponding temperature is called Kraft temperature (T_K). At this temperature, the micelles are formed and the solubility is significantly increased and equal to the surfactant's CMC at the same temperature. Actually, micelles form only above the CMC and above the Kraft temperature. Below the Kraft temperature, the maximum solubility of the surfactant will be lower than the CMC, meaning micelles will not form. Kraft temperature increases with a long hydrophobic carbon chain as its solubility in water decreases. Meanwhile, the Kraft points have no relationship with a series of methylene linking groups. The Kraft point of the Gemini surfactant containing benzyl is 1°C higher than those containing methylene series, although the hydrophobic carbon chain is the same in both cases [52]. Usually, electrolytes raise the Kraft temperature, whereas there is no general trend for the dependence on counter ions. However, the Kraft point is typically much higher in presence of divalent counter ions than monovalent counter ions. For cationic surfactants, the Kraft point is typically higher for bromides than chlorides, and still higher for the iodides.

6.4 Pressure

Pressure also has a significant effect on the self-organization of surfactants. Pressure initially retards the association and after a certain limiting value, known as threshold value (100–200 MPa), the process is favored. The release of surfactant

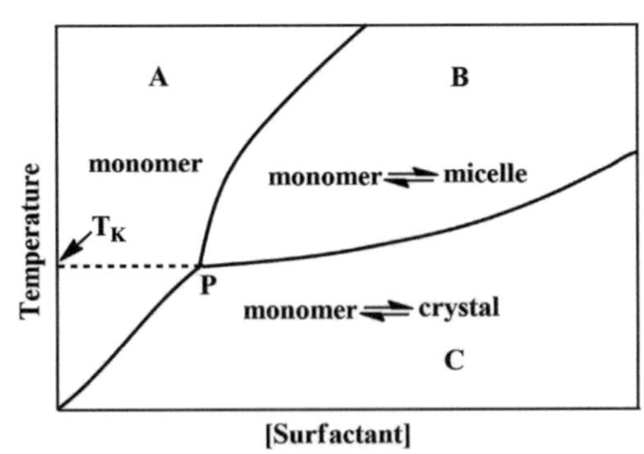

Figure 9.
The schematic representation of the different phases of surfactant at different temperatures and T_K represents a 'Kraft temperature'.

monomers from the micelles in the lower range of pressure and their association at higher pressure together with the changed dielectric constant of the solution play the specific roles in surfactant organization. This has been supported by the measurement of aggregation number, which shows a minimum for ionic surfactants and a rapid initial decrease for non-ionic surfactants with respect to pressure. The effect of pressure on CMC for ionic surfactant solutions is easily determined from the conductivity measurements of the solution at elevated pressures [53]. On the other hand, no direct experimental determination of the effect of pressure on CMC for non-ionic surfactant solutions has been reported, except for high-pressure NMR techniques. For example, ^{13}C-NMR experiments have been reported in the literature [54] to obtain CMC and micelle aggregation numbers for non-ionic surfactant solutions at ambient pressure. The CMC's for non-ionic surfactant solutions at elevated pressures have been calculated, however, based on measured aggregation numbers at ambient pressure and estimated hydrocarbon compressibilities in the micelle core. These calculations indicate that the cmc increases with increasing pressure [55].

6.5 Presence of additives

Additives may have significant effects on surfactant self-organization [56, 57]. A salting out effect of salts influencing the surfactant activity to assist easier aggregation may arise. The study of salt effects is important as amphiphiles are mostly handled in electrolyte environments in chemical research. Non-electrolytes may both increase and decrease micellization tendency of surfactants. The matter is complicated because additives can affect solvent structure and polarity and interact directly with surfactants. Urea and guanidine hydrochloride are obvious in this case, they greatly inhibit micellization and can break the water structure [57].

7. Determination of CMC

There are various techniques/methods have been frequently used for determination of CMC values such as Conductometric, tensiometry, cyclic voltammetry, densiometry, dye solubilization, UV–visible and fluorescence spectroscopy etc. Different mathematical approaches have also been applied to determine the CMC values from the raw experimental data. In this chapter, some techniques are discussed briefly.

7.1 Conductometric

Conductivity is a common technique to determine CMC for ionic surfactants, which behave as electrolytes in water. This technique cannot be used for non-ionic surfactants since these surfactants have a negligible effect on the conductivity of the solution. Conductivity measurements are usually carried out using a digital conductivity meter (typically cell constant 1.0 cm^{-1}) of surfactant solution in water. The conductivity meter is calibrated using aqueous potassium chloride (KCl) solutions in the proper concentration range before measurement. The measured conductivity values have been multiplied with cell constant to obtain the specific conductivity (κ, kappa) values.

Two linear segments with different slopes can be determined and identified to the monomeric and micellar states of surfactants in solution. In particular, the increase in

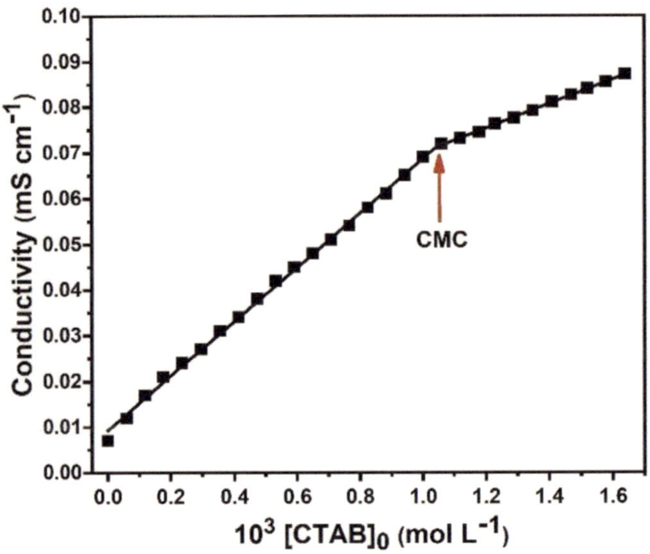

Figure 10.
Variation of conductance with different concentrations of CTAB.

conductivity with concentration is greater when the surfactants in solution are present as unimers than in the presence of micelles [34, 58]. CMC can be calculated from the breakpoint of specific conductivity versus concentration curve shown in **Figure 10**.

7.2 Tensiometry

The surface tension of surfactants at various concentrations is measured by using tensiometer. The tensiometer should be properly cleaned before the measurement and the reliability of the instrument is verified by measuring the surface tension of pure water. The value of pure water should be 71.97 mN/m at 298.15 K. Typically, a surfactant stock solution is added to pure water, the solution is stirred for 1 min before the surface tension is measured, and then the reading is taken. The surface tension is linearly dependent on the logarithm's surfactant concentrations [30, 32]. However, above the CMC, the surface tension is independent of the surfactant's concentration. Mostly, CMC is the point of intersection between two slopes. The intersection points are deteriorations of the straight line of the linear dependent region and a straight line passing through the flat terrain where no change in surface tension as shown in **Figure 11**.

7.3 Fluorometry

Fluorescence measurements are taken using a fluorimeter of a 10 mm path length quartz cuvette. A fluorescence probe (pyrene) is used in very low concentration around 2 μM. In this method 1st excited the probe in its excitation wavelength and then measure the emission spectra. The slit widths are kept in fixed for excitation and emission.

The fluorescence spectra of pyrene is shown in **Figure 12a** in presence of AOT. The spectra showed five emission peaks at 375 (I_1), 380 (I_2), 385 (I_3), 390 (I_4) and 395 (I_5)

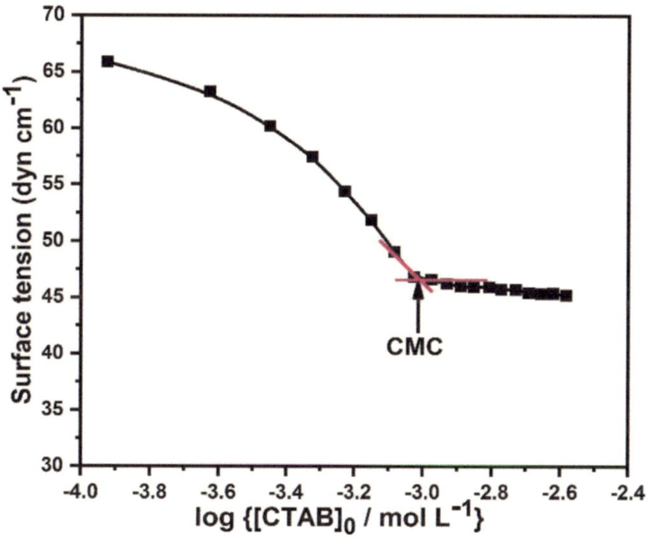

Figure 11.
The observed surface tension vs. concentration profile for CTAB.

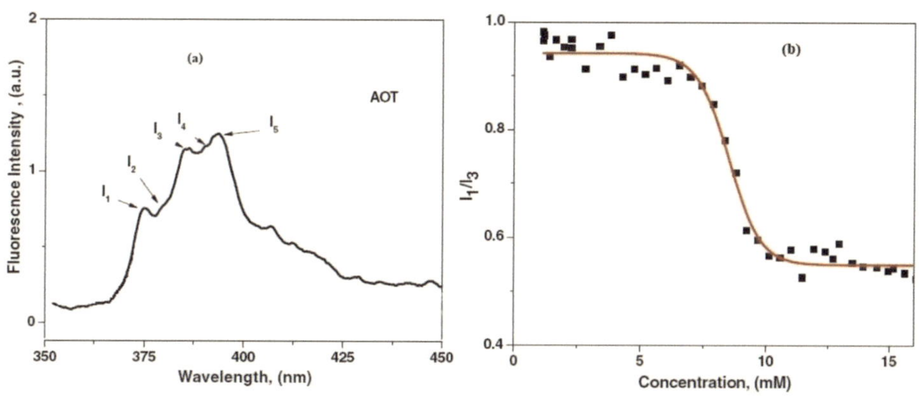

Figure 12.
(a) Fluorescence spectra of pyrene in presence of AOT in MetOH (b) I_1/I_3 vs. AOT concentration profile for pyrene. The sigmoid line corresponds to the Sigmoid-Boltzmann equation (SBE) fitting.

nm for pyrene [58, 59]. In polar solvents i.e. water, MeOH etc., pyrene shows high fluorescence intensity for emission peaks I_1 and I_3. Below the CMC (absence of micelles), pyrene senses the polar environment of methanol (MeOH) molecules. Consequently, the ratio of fluorescence emission intensities corresponding to the first and third vibrational peaks (I_1/I_3) is high. But above the CMC (presence of micelles), pyrene molecules are solubilized in the interior micellar phase due to high hydrophobicity of pyrene. This is a non-polar like solvent, so the environment sensed by pyrene is less polar. Therefore, the ratio I_1/I_3 decreases (**Figure 12b**). Such a decrease indicates that the microenvironment around fluorescent probe changes with surfactant concentrations becoming more hydrophobic, because of probe (pyrene) interactions with the surfactant micelles. Similar sigmoid shaped curve have been obtained in presence of SDS, SDDS (N-lauroyl sarcosine sodium salt) [60].

7.4 Refractive index

In refractive index experiment, the CMC of surfactant is determined by the following method. Firstly, a concentrated surfactant solution is added over time to a fixed volume of distilled water in a beaker with an automatic burette. Secondly, the solution in the beaker is stirred using a magnetic stirrer. The control program automatically measures the refractive index of the sample at regular time intervals (typically 3–5 s) with the addition of surfactant solution. A plot of refractive index versus concentration is also displayed in the program window. In fact, the automatic burette is operated at a constant speed and the time axis of the chart can be converted to concentration units. The CMC of the surfactant is measured by a breakpoint from the plot under investigation at constant temperature.

The working principle of CMC detection using the refractive index depends on two factors: (i) higher refractive index of micelles than monomers and (ii) surfactant molecules adsorption on the sensing tip. The hydrophobic fiber core is made of silica and the bare core in the sensing region is immersed in the sample solution. The surfactant molecules begin to aggregate at the air–solution interface with increasing concentration. The hydrophilic parts of the surfactant molecules begin to interact with the hydrophilic surface of the sensing tips above the CMC [61]. A clear indication of CMC at this point can be obtained due to the combined effects of micelle formation and surfactant adsorption that increases the refractive index more rapidly with concentration. **Figure 13** is the plot of concentration vs. refractive index to depict the CMC of SDS. The CMC value is determined from the intersection of the lines above and below the breakpoint.

7.5 Viscometry

The critical micelle concentration of the surfactant is determined by measuring viscosity as a function of surfactant concentration using the viscometer by the Poiseuille equation,

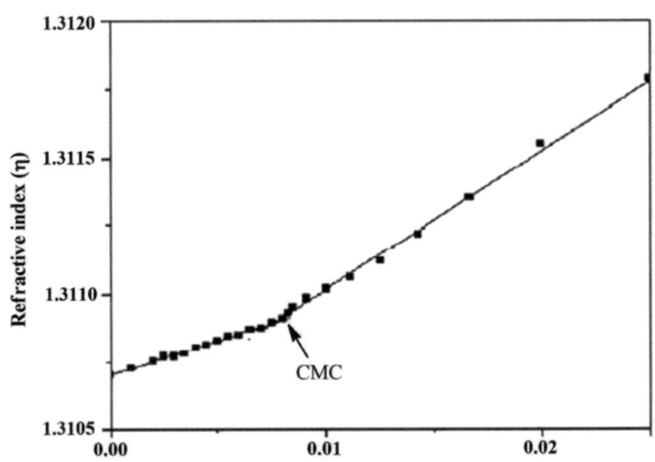

Figure 13.
Plot of refractive index versus concentration of SDS at 298 K. The intersection of the breakpoint indicates the CMC of SDS.

$$\eta = \frac{\pi \Delta \mathrm{P} r^4}{8 l \mathrm{V}} \Delta t$$

Whereas ΔP denotes the pressure in tube, r is tube radial, l is tube length and V is tube volume of a viscometer. All these parameters are kept constant during the measurements. Δt is the flow time. The viscosity of the pure solvent (η_0) and the viscosity of the surfactant solution (η_s) can be estimated by the flow time ratio,

$$\frac{\eta_0}{\eta_s} = \frac{\Delta t_s}{\Delta t_0}$$

The flow time of the solution containing surfactant increases relative to that of a pure solvent and hence the viscosity increases. The concentration of the surfactants is plotted against the estimated viscosity shown in **Figure 14**.

From **Figure 14**, it is clear that there is a sharp change in the linear viscosity increase in both surfactant systems, AOT and SDS [59]. The slope of the plot line changes after a critical point. A critical point is initially reached, with a gradual increase in the surfactant concentration of the dispersion. Afterwards, the surfactant molecules redistribute into the solution to form micelles. The viscosity of the solution increases sharply in micelles as the micellization process leads to a more compact solution structure. It is then possible to estimate the CMC from the intersection of the two straight lines in **Figure 14**.

7.6 Density measurements

Density is an important physical parameter of solutions that varies with the state of aggregation of surfactants. In particular, the increase in the density of a monomer solution of the surfactant per unit mass is greater than its aggregation state where micelles are present [62]. It is related to the different volume fractions of monomers and micelles in solution. Monomers have a higher volume fraction than micelles due to

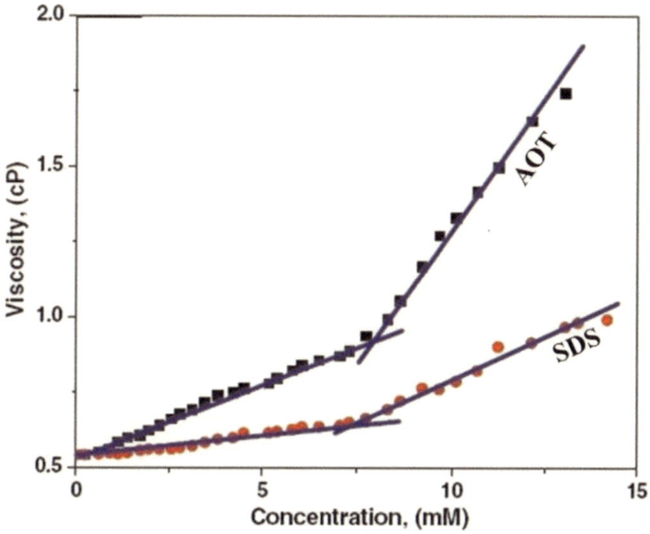

Figure 14.
Plot of viscosity vs. concentration of the surfactant.

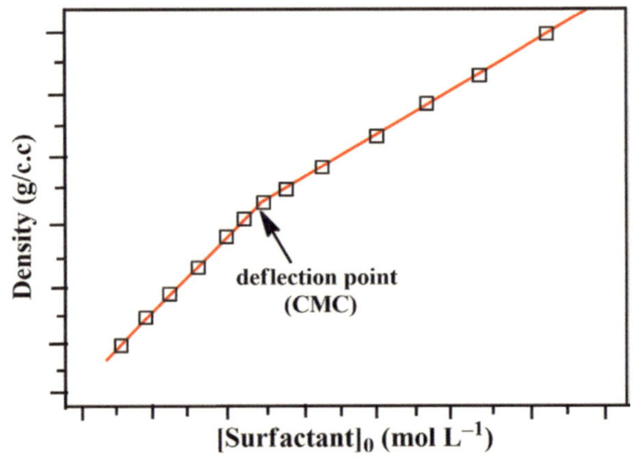

Figure 15.
Schematic representation of density versus surfactant concentration, the defection point shows the CMC of the surfactant.

higher hydration. Therefore, water is more bounded in the presence of monomers than micelles. Micellization, indeed, is a dehydration process. It releases more free water than bound water. Consequently, the increase in volume per unit mass of the surfactant as monomers is less than that of micelles.

The CMC of the surfactant can be determined from the deflection point [62] of the density versus concentration plot shown in **Figure 15**. The extent of the increase in density in the monomer state (slope) depends on the different grade of hydration and the molecular weight of the surfactant.

However, the density measurement technique is not suitable for CMC determination of non-ionic surfactants because the density variation is not appreciable with low surfactant concentrations and becomes comparable to pure water.

8. Different models for micellar catalyst reactions

Several models have been developed to explain the surfactant effects on reaction rates, both pre-micellar and post-micellar. A few models are briefly discussed in this section.

8.1 Merger and Portnoy model (/simple distribution model)

Merger and Portnoy [63, 64] proposed the first kinetic model, which related micelles as enzyme like particles and successfully fitted inhibited bimolecular reactions like basic hydrolysis of p-nitrophenyl acetate, mono-p-nitrophenyl dodecanedioate and p-nitrophenyl octanoate in presence of laurate micelle.

8.1.1 The fundamental concepts and assumptions of this model

In a homogeneous surfactant solution (above the critical micelle concentration), the reactive site of a substrate may exist in one or more of the following environments: the micelle interior, the micelle-water interface, and the bulk solvent. The nature of

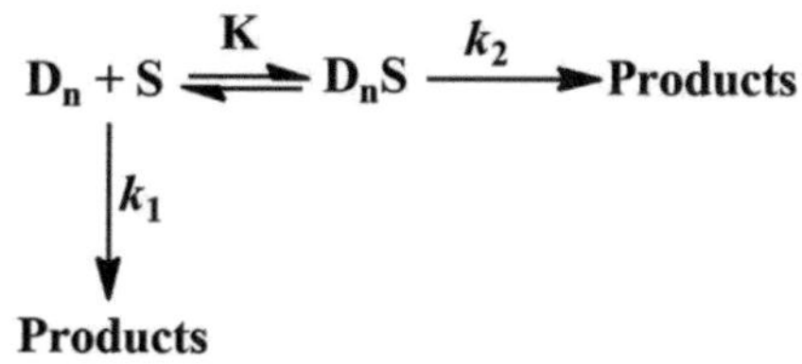

Figure 16.
Different steps in Merger and Portnoy model.

the micelle interior, formed by the lyophobic portion of the surfactant, is very complicated. From high-resolution NMR experiments, it appears that (i) the centre of the micelle resembles a liquid hydrocarbon and (ii) water can penetrate the micelle so that part of the alkyl chain (possibly the first five carbons from the ionic group) is exposed to the solvent. Based on different site of micellar phase, Menger and Portnoy in their model have been assumed that the variation of the rate with surfactant concentration depends on the distribution of substrate between aqueous and micellar pseudophase and they proposed the following **Figure 16**.

Where, D_n, S and D_nS represent micellar surfactant (/detergent), free substrate (e.g. ester) and adsorbed substrate respectively. k_1 and k_2 are the first order rate constants for the reaction in the aqueous and micellar pseudophase, respectively. In order to evaluate the micelle concentration, they make use of the 'phase separation' concept, which assumes that the unassociated surfactant concentration remains constant above the CMC.

If the average number of molecules per surfactant micelle is n (i.e., aggregation number of the micelle), then the micelle concentration is approximated by Eq. (1)

$$[D_n] = \frac{[D]_T - CMC}{n} \tag{1}$$

Where, $[D]_T$ signifies the total concentration of surfactant and $[D_n]$ is the micellized concentration of surfactant. If the average number of molecules per laurate micelle is 33 [64] then the micelle concentration is approximated by Eq. (1) as

$$[D_n] = \frac{[\text{laurate}]_T - CMC}{33}$$

The rate constant, k_2 can be now determined in the following way:

The binding constant (K) according to **Figure 16** is given by

$$K = \frac{[D_nS]}{[D_n]\ [S]_{free}} \tag{2}$$

(Or,) $[D_nS] = K[D_n]\ [S]_{free}$and K expressed in terms of the concentration of micelle rather than micellized surfactant.

Now, the initial concentration of substrate written as

$$\begin{aligned} [S]_0 &= [S]_{free} + [D_nS] \\ &= [S]_{free} + K[D_n]\ [S]_{free}\ (\text{substitute } [D_nS] \text{ from eq.2}) \\ &= (1 + K[D_n])\ [S]_{free} \end{aligned}$$

$$\therefore [S]_{free} = \frac{[S]_0}{1 + K[D_n]} \tag{3}$$

Again, rate of reaction,

$$\begin{aligned} r &= k_1[S]_{free} + k_2[D_nS] \\ &= k_1[S]_{free} + k_2K[D_n][S]_{free} \\ &= (k_1 + k_2K[D_n])[S]_{free} \end{aligned}$$

Substituting [S]free from Eq. (3), we get,

$$r = (k_1 + k_2K[D_n])\frac{[S]_0}{1 + K[D_n]} = k_{obs}[S]_0 \tag{4}$$

Where,

$$k_{obs} = \frac{(k_1 + k_2K[D_n])}{(1 + K[D_n])} \tag{5}$$

Now,

$$k_{obs} - k_1 = \frac{(k_1 + k_2K[D_n])}{(1 + K[D_n])} - k_1 = \frac{(k_1 + k_2)K[D_n]}{1 + K[D_n]}$$

$$\frac{1}{(k_{obs} - k_1)} = \frac{1}{(k_2 - k_1)} + \frac{1}{K(k_2 - k_1)} \times \frac{1}{[D_n]} \tag{6}$$

Eq. (6) is very useful to determination of both k_2 and K from the linear plot of 1/ $(k_{obs} - k_1)$ vs. $1/[D_n]$. Its linearity is remarkable in view of the assumptions made in this analysis: (i) substrate and surfactant monomer do not form complex; (ii) the micellation process does not perturbed by substrate; (iii) substrate and micelle associate with a 1: 1 stoichiometry; (iv) micellation happens exactly at CMC rather than in a small concentration range; (v) Eq. (1) is valid.

Eq. (5) has a form similar to the Michaelis–Menten equation of enzyme kinetics; although enzyme reactions are generally followed with substrate in excess over the enzyme whereas in the normal reaction generally the surfactant is in large excess over the substrate. Eq. (6), which is similar to the Lineweaver-Burk equation of enzyme kinetics.

Eq. (6) was first applied to the inhibition of the saponification of p-nitrophenyl alkanoates by micelles of sodium laurate [64] and was subsequently applied to unimolecular hydrolyses of dinitrophenyl phosphate dianions and 2,4-dinitrophenyl sulphate monoanion catalyzed by cationic micelles. These observations suggest that the model is useful in analyzing micellar catalysis and inhibition. There are few reactions (**Figure 17**), whose kinetic data satisfied the above rate expression as well as the model [65, 66].

8.1.2 Merits and demerits of this model

The simple distribution model (Eqs. (5) and (6)) adequately treats micellar effects upon unimolecular reactions and inhibited bimolecular reactions because these micellar effects can be treated in terms of the distribution of only one reactant between the

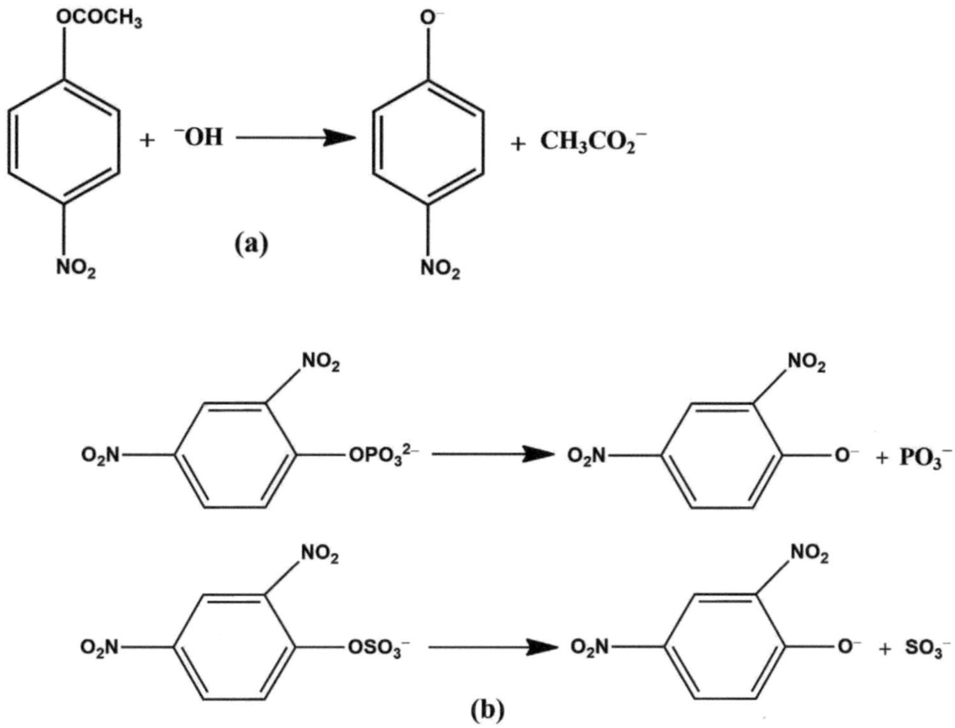

Figure 17.
(a) Micellar inhibition of bimolecular reaction; (b) cationic micellar catalysis of unimolecular reactions.

aqueous and micellar pseudophases. This model also fitted spontaneous reactions [67, 68]. However, this simple model could not simulate the rate maxima characteristic of micellar-enhanced bimolecular reactions, where rate constants generally go through maxima with increasing surfactant concentrations, because the distribution of two reactants must be considered, and this problem is considered in the next model. Berezin and co-workers [68] developed the first general treatment based on the pseudophase model and successfully simulated spontaneous and bimolecular reactions between neutral organic reactants. Later the pseudophase model was modified to simulate bimolecular, ion-molecule, reactions.

8.2 Pseudophase (PP) Model (or Berezin Model)

Berezin and co-workers [68] developed the first general treatment based on the pseudophase model and successfully simulated spontaneous and bimolecular reactions between non-ionic reactants and with partial success to reactions between neutral organic substrates and hydrophobic anionic nucleophiles. All pseudophase kinetic models are supported by two common assumptions. These are: (i) the micelles and water are different phases and (ii) the rate constants depending on the distribution of substrate (S) and nucleophile (N) in the micelles and aqueous phases in addition to surfactant and salt. In a few studies, it is also assumed that changes in micelle size and shape are not very important so that only those factors, which control the distribution of reactants, will significantly affect the observed reaction rate. The model correctly predicts the rate maxima, which are typically observed with micellar catalyzed bimolecular reactions. Micelle formation is a highly cooperative phenomenon. When the surfactant concentration exceeds the critical micelle concentration (CMC), all additional surfactant forms micelles. Thus, the micellized surfactant can be defined as follows:

$$[D_n] = [D]_T\text{–}CMC \tag{7}$$

Where, $[D]_T$ is the stoichiometric surfactant (/detergent) concentration and the CMC is that obtained under experimental reaction conditions. Often $[D]_T >> CMC$ and the CMC can be neglected. According to this model, the micellized surfactant and the solute are in thermal equilibrium throughout the reaction. The second-order rate constants (k_2^{w} and k_2^{m}) can be considered as the sum of the concurrent reaction rates in each pseudophase shown in **Figure 18**. In **Figure 18**, subscripts m and w indicate the micellar and aqueous pseudophases, respectively.

Micellar binding of substrate is governed by both columbic and hydrophobic interactions and is generally described by a binding constant, K_s:

$$K_S = \frac{[S_m]}{[S_w]\,[[D_n]} \tag{8}$$

Values for K_s can usually be estimated, in the absence of N, by spectroscopy, solubility, liquid chromatography, or ultrafiltration, and they increase with increasing substrate hydrophobicity (R is the length of alkyl chain of the substrate). Spectroscopically it is shown that the bound polar substrates are on the micellar surface rather than in the hydrophobic core. The rate expression can be obtained from **Figure 18** using Eq. (9) as

$$k_{obs} = \frac{k_2^{w}[N_w] + k_2^{m}K_sN_m[D_n]}{1 + K_s[D_n]} \tag{9}$$

Where, the term N_m is the local molar concentration of the ionic reactant in the micellar pseudophase and can be written as Eq. (10).

$$N_m = \frac{[N_m]}{[D_n]\,V_m} \tag{10}$$

Similarly, V_m is the molar volume of the reactive region and the micellar fractional volume where the reaction occurs is denoted as $[D_n] \times V_m$. In the aqueous phase, the

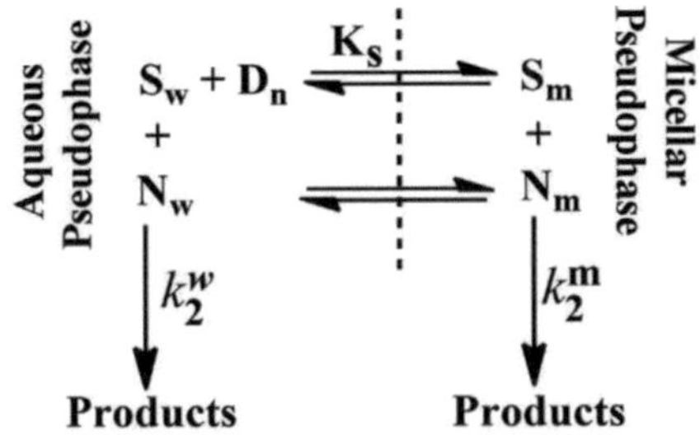

Figure 18.
Different steps in Pseudophase model.

concentration of ions and molecules is usually expressed in molarity of the total solution volume since the micellar pseudophase volume exceeds about 3% (for about 0.1 M surfactant) of the total solution volume.

The relative proportions of the overall reaction occurring in the aqueous and micellar pseudophases are dependent on both $[D_n]$ and K_s [3, 4, 69, 70] as indicated in Eq. (8). Also, Eq. (9) illustrates a fundamental property of all pseudophase models, that reaction rate within the micellar pseudophase depends upon the local concentration of N within the micellar pseudophase and not its stoichiometric concentration.

8.2.1 Demerits of the oseudophase model

The pseudophase model for the micellar-catalyzed bimolecular reactions is based on various assumptions about the interactions of solutes with micelles and the appropriate units of concentration. However, there are some questions about the validity of assumptions about whether the reaction occurs in the aqueous pseudophase or in micelles. It is also not clear whether a reactant in water attacks another reactant in the micelle. Furthermore, how are the overall rate constants in the two pseudo-phases so generalized in Eqs. (5) and (9)? It also appears that this generally accepted hypothesis fails in certain special cases.

The model accurately predicts the observed maximum rate of micellar catalyzed bimolecular reactions. However, it has not been thoroughly tested for micellar catalyzed bimolecular reactions involving hydrophilic ions, with the exception of reactions involving hydrogen ions and nucleophilic addition to carbocations.

The pseudophase model explicitly assumes that changes in micelle size and shape are not very important, so that only those factors, which control the distribution of reactants, will significantly affect the observed reaction rate. Buffer effects are particularly difficult to understand, so the model applicable to the selected reactions where would not require use of buffers. Finally, the presence of both reactive and unreactive counter ions results in uncontrolled variations in distribution of them between the micellar and aqueous pseudo phases. Thus, it is further complicated to interpret the CMC.

8.3 Pseudophase ion exchange (PIE) model

The treatment of micellar catalysis and inhibition is based on the assumption that reaction occurs in the micellar and aqueous pseudo-phases and that equilibrium is maintained between reactants in the two pseudo-phases. The binding constant is given by Eq. (8) and the corresponding rate constant is given by Eq. (9). Now the problem is nonreactive counter ions. A widely used approach is to assume that counter ions bind to a micelle according to an ion-exchange equation

$$N_m + X_w \rightleftharpoons N_w + X_m$$

A useful physical picture for interpreting counter ion effects is based on Stigter's model, which interprets the effect of added salt on the surface potential of micelles [71]. The counter ions are supposed to be distributed in two discrete sections of an ionic bilayer. These are (i) the Stern layer strongly bound with the counter ions that move with the micellar aggregate (the kinetic micelle), and (ii) the Gouy-Chapman layer that is loosely bound with the remaining counter ions according to the Boltzmann distribution law in the aqueous phase (**Figure 19**). Similar double-layer

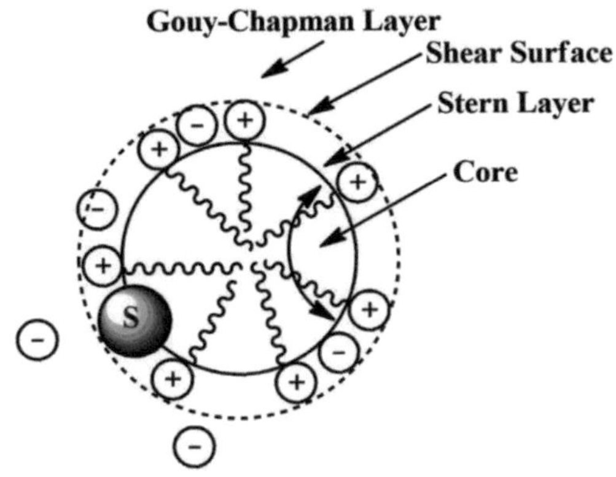

Figure 19.
Model of a hypothetical cationic micelle showing the locations of head groups, surfactant chains, and counter ions.

models describe counter ion distributions around planar interfaces, charged electrodes, and lyophobic colloids. Since the potential drop across the Stern layer is relatively insensitive to increasing the concentration of counter ions, the counter ions concentration in the Stern layer and the fraction of counter ions bound to the micelle surface will be approximately constant (β) [72, 73]. The theoretical concept of Stigter enables the calculation of the value of β [72, 73].

According to **Figure 19**, N is a counter ion to ionic micelles; binding substrate, S brings it into a microenvironment. Consequently, the local concentration of N increases and hence increases the observed rate constant (k_{obs}). Again, when N is a co-ion, binding S brings it into a microenvironment with a much lower concentration of N, which reduces k_{obs}. Two assumptions are considered to estimate the concentration of N in micelles in the PIE model. The assumptions consist of (i) the micellar surface, known as a selective ion exchanger, and a relationship between inert counter ions (X), and reactive counter ions (N), given by Eq. (11).

$$K_N^X = \frac{[N_w]\,[X_m]}{[N_m]\,[X_w]} \tag{11}$$

(ii) the fraction of the surface occupied by the two counter ions, X and N is constant and given by the degree of counter ion binding, β, expressed in Eq. (12).

$$\beta = \frac{[N_m] + [X_m]}{[D_n]} \tag{12}$$

The value of β varies with the variety of counter ions and head groups, and the experimental values are around 0.6–0.9 [72, 73]. This can be estimated from the fractional micellar charge a ($a = 1 - \beta$). β is usually insensitive to surfactant and salt concentration, indicating significant specific interactions at the micelle surface. Thus if $[N_m]$ and $[N_w]$ are calculated from Eqs. (11) and (12), the variation of k_{obs} with $[D_n]$ (Eq. 9) can be used to estimatek_2^m. In practice, non-linear simulation in computer base is usually used to fit the variation of k_{obs} with $[D_n]$ in terms of the various parameters in Eq. (9). However, the above ion-exchange pseudophase model based on Eq. (11) involves the assumption that β does not change as reactive counter ion, e.g., N is added or the surfactant concentration changed. This assumption appears to fail when the only anions in the solution are very hydrophilic, e.g., OH^- and F^-. In these systems the kinetic data can be fitted by a mass-action model in which assumes that β increases with increasing $[OH^-]$ or $[F^-]$ in Eq. (13). β appears to increase significantly, with their total concentrations indicating that they interact mostly coulombically and that appreciable fractions of these ions are in the diffuse double layer.

$$D_n + OH^-_w \overset{K_{OH}}{\rightleftarrows} OH^-_m \tag{13}$$

The binding of inert counter ion, e.g., Br^-, could similarly follow Eq. (14)

$$D_n + Br^-_w \overset{K_{Br}}{\rightleftarrows} Br^-_m \tag{14}$$

Thus,

$$K_{OH} = \frac{[OH^-_m]}{([D_n] - [OH^-_m] - [Br^-_m])([OH^-_T] - [OH^-_m])} \tag{15}$$

$$\text{and, } K_{Br} = \frac{[Br^-_m]}{([D_n] - [OH^-_m] - [Br^-_m])([Br^-_T] - [Br^-_m])} \tag{16}$$

OH^- ion appears to bind very weakly to cationic micelles so that in presence of CTAB, $[OH^-_T] > > [OH^-_m]$. With this assumption and Eqs. (12) and (13), and mass balance, the value of $[OH^-_m]$, is given by

$$\begin{aligned}(K_{Br} + K_{Br}K_{OH}[OH^-_T])[OH^-_m]^2 + (1 + K_{OH}[OH^-_T] + K_{Br}[Br^-_T] - K_{Br}[D_n])(K_{OH}[OH^-_T])[OH^-_m] \\ - K_{OH}[OH^-_T][D_n] \\ = 0\end{aligned} \tag{17}$$

The quadratic Eq. (17) can be solved for given concentrations of OH^- and CTAB, assuming that K_{OH} and K_{Br} are independent parameters and also assuming their values. Using the calculated value of $[OH^-_m]$ from Eq. (17), the observed rate expression becomes

$$k_{obs} = \frac{k_2^w[OH^-_w] + k_2^m K_S[OH^-_m][D_n]}{1 + K_S[D_n]}$$

This treatment makes no assumptions regarding the value of β, which in principle can range from 0 to 1. Both the ion-exchange model in Eq. (11) and the mass action model in Eqs. (13) and (14) are limiting models, which fit the kinetic data, but only approximate to reality.

8.4 Piszkiewicz co-operativity model

There are important similarities between micelle-catalyzed reactions and enzyme-catalyzed reactions. The structures of both micelles and enzymes are similar in that they have hydrophobic cores with polar groups on their surfaces. The structures of micelles are disrupted by common protein denaturing agents such as urea and guanidinium salts. Both catalytic micelles and enzymes bind substrates in a noncovalent manner. The kinetics of micellar catalysis resemble that of enzymatic catalysis in that the micelle may be saturated by the substrate; and, conversely, the substrate may be saturated by the micelle [74].

Piszkiewicz showed further similarities in micellar and enzyme reactions [75–77]. The sigmoid curves are obtained by plotting the rate constants of micelle-catalyzed reactions with respect to the detergent concentration. This behavior is referred to as positive cooperativity (or positive homotropic interactions) of enzyme catalyzed reactions. The Hill model, a kinetic model [78], also describes the sigmoid dependence of rate on substrate concentration. Piszkiewicz in his model demonstrated that micelle-catalyzed reactions also are models of enzyme-catalyzed reactions that show positive cooperativity [4, 77]. In most cases, the rate of micellar reaction is much faster than that of other common observations. The micelle-catalyzed reactions occur in two steps. The first step is the complex formation of detergent-substrate (DnS) which is the fast step and the other is the catalytic step, which is the rate limiting step. Therefore, two rate constants (k_w and k_m) are required to describe these reactions. The terms, k_w and k_m, denotes the rate constants in absence and presence of detergent respectively and k_m/k_w is equal to the rate acceleration effected by micellar catalysis.

However, this model assumes that a substrate, S, and a number of detergent molecules (nD), aggregate to form catalytic micelles, D_nS, which may then react to yield product shown in **Figure 20**.

K_D is the dissociation constant of the detergent-substrate (DnS) complex towards its free components. k_w and k_m are the rate constants in absence and presence of detergent. Indeed, there are several steps and sequential equilibrium steps involved in the aggregation of detergent and substrate molecules to form the catalytic micelle, D_nS. For convenience, the steps are presented as a single association step (**Figure 20**) in this article. For the reaction indicated in **Figure 20**, the observed rate constant is expressed as a function of the concentration of detergent by Eq. (18).

$$k_{obs} = \frac{k_m[D]^n + k_w K_D}{K_D + [D]^n} \tag{18}$$

This equation may be rearranged and its logarithmic form can be written as

$$\log\left(\frac{k_{obs} - k_w}{k_m - k_{obs}}\right) = n \log [D] - \log K_D \tag{19}$$

$$nD + S \underset{K_D}{\rightleftharpoons} D_nS$$

$$D_nS \xrightarrow{k_m} \text{Product}$$

$$S \xrightarrow{k_w} \text{Product}$$

Figure 20.
Reaction steps involved in the Piszkiewicz's model.

From Eq. (19) above, a linear curve having a slope of n is obtained by plotting of log $[(k_{obs} - k_w)/(k_m - k_{obs})]$ vs. log [D] for a micelle-catalyzed reaction. When log $[(k_{obs} - k_w)/(k_m - k_{obs})] = 0$, n log[D] = log K_D. Also, at log $[(k_{obs} - k_w)/(k_m - k_{obs})] = 0$, catalysis by detergent shows one-half of its maximum effect on rate constant i.e. the log of the detergent concentration at which half-maximal velocity is obtained. For convenience, the value of log [D] at this point is designated as log$[D]_{50}$, and it is equal to (log K_D)/n.

8.4.1 *Effect of temperature variation and detergent structure*

Several experiments show that the value of n depends on temperature. The average value of n appears to increase as temperature decreases. E. H. Cordes and his co-worker [79] were determined an average value of n of 1.87 at 40°C for the hydrolysis of methyl orthobenzoate in presence of sixteen different detergents. In contrast, the hydrolysis of the substrate catalyzed by six detergents at 25°C had an average 'n' value of 3.48. However, three detergents with an average 'n' value of 1.41 inhibited the hydrolysis of methyl orthobenzoate [77]. Only a few are listed in **Table 3**.

8.4.2 *Effect of changes of substrate structure*

Several series of reactions have been studied in which a single detergent was employed within the series, but slight variations in the structure of the substrate were made. One such series of reactions were reported by Dunlap et al. [80] in which hydrolyses of para-substituted benzaldehyde diethyl acetals occur in presence of sodium dodecyl sulfate. **Table 4** summarizes the constant values, log $[D]_{50}$ and n, for

Detergents	Temp. (°C)	No. of k_{obs} values used	Log $[D]_{50}$	n
Sodium octyl sulfate	25	4	−0.91	3.30
Sodium decyl sulfate	25	3	−1.42	5.46
Sodium dodecyl sulfate	25	8	−1.96	3.67
Sodium dodecyl sulfate	40	4	−1.90	5.86
3-Hexadecyl sodium sulfate	40	5	−2.45	1.55
5-Hexadecyl sodium sulfate	40	5	−2.35	1.82
Sodium 2-hexadecyloxy-1-methylethyl sulfate	40	7	−2.69	1.63
Disodium 2-sulfo-2-methyloctadecanoate	40	4	−2.52	2.14
Sodium methyl-α-sulfostearate	40	4	−2.56	1.43
Sodium 2-dodecylbenzenesulfonate	40	4	−2.49	1.75
Dimethyldodecylammonium propanesulfonate[a]	25	6	−1.68	1.03
Dimetliyldodecylphosphine oxide[a]	25	4	−1.90	1.38
Dimetliyldodecylammonium acetate[a]	25	3	−1.85	1.82

[a]*Hydrolysis of methyl orthobenzoate was inhibited by these detergents [80].*

Table 3.
Observed rate constants and values of n for the hydrolysis of methyl Orthobenzoate in the presence of various detergents.

Substituents	σ	No. of k_{obs} values used	Log[D]$_{50}$	n
p-NO_2	0.710	2	−2.06	4.47
p-Cl	0.373	4	−2.03	5.02
p-F	0.337	5	−1.96	3.74
p-OCH_3	0.115	4	−2.01	4.52
p-H	0.00	4	−1.98	5.41
p-CH_3	−0.069	5	−1.99	4.33

Table 4.
Derived values of log $[D]_{50}$ and n for hydrolyses of Para-substituted benzaldehyde diethyl acetals in presence of sodium dodecyl sulfate.

these reactions. It is clear from the data shown in **Table 4** that the value of log $[D]_{50}$ remains virtually constant as the inductive substituent constant (σ) of the para substituent is varied from −0.069 for p-methyl to +0.710 for p-nitro. Also, show that there is no consistent variation of n with σ.

A somewhat different situation is seen for the solvolyses of several esters by the nucleophilic detergent p-trimethylammonio benzyl decylamine chloride as reported by Bruice et al. [81]. **Table 5** summarizes the parameters, log $[D]_{50}$ and n, for these reactions are clearly a function of the structure of the substrate. Generalized data shown in **Table 5** can be explained in different ways. Firstly, log $[D]_{50}$ increases as the chain length of the carboxylic acid portion of the negatively charged ester increases. It is clear from the result that the better binding of the longer chain substrates within the catalytic micelle complex (D_nS) due to hydrophobic interactions. Secondly, 'n' appears to decrease as the chain length of the carboxylic acid moiety increases, though there are few exceptions. In this reaction, the substrate has a much longer aliphatic chain than does the detergent and thus this is different from other series of reactions. This structural difference probably radically changes the mechanism of the formation of the catalytic micelle (D_nS).The catalytically active complex may form firstly the aggregation of substrate molecules then include the nucleophilic detergent molecules.

Esters	No. of k_{obs} values used	log[D]$_{50}$	n
3-Nitro-4-acetoxy benzenesulfonate	7	−2.34	4.14
3-Nitro-4-hexanoyloxy benzenesulfonate	6	−2.69	3.01
3-Nitro-4-octanoyloxy benzenesulfonate	5	−3.05	1.63
3-Nitro-4-decanoyloxy benzenesulfonate	5	−3.03	1.25
3-Nitro-4- hexadecanoyloxy benzenesulfonate	4	−2.76	4.87
3-Nitro-4-acetoxyphenyltrimethylammonium iodide	10	−2.38	1.91
3-Nitro-4-octanoyloxyphenyltrimethylammonium iodide	9	−2.03	1.80
o-Nitrophenyl acetate	10	−1.92	1.58

Table 5.
Nucleophilic solvolyses of different esters derived from p-*Trimethylammonio benzyl decylamine chloride and n values in micelle.*

From observing the effects of changing substrate structure, it is clear that changing log $[D]_{50}$ or n requires a major structural change. The change in electronic inductive effects may not be sufficient. Moreover, the relation between the nature of changes in log $[D]_{50}$ and n is a complex function of the nature of structural changes.

8.4.3 Limitations

The use of *log* $[(k_{obs} - k_w)/(k_m - k_{obs})]$ causes significant scale expansion at both ends, but very little distortion in the mid-range. In the range where the rate constants to be subtracted are about an order of magnitude from each other, a reasonable value of n can be predicted. However, the precision of linear plot points outside this range degrades sharply. This degradation of precision at the extremes may contribute to the lack of precision in the value of n. The term, log K_D is equal to n log $[D]_{50}$. As n log $[D]_{50}$ is the product of two terms, it reflects the uncertainty in both, K_D and n. Moreover, the K_D units vary with n reaction to reaction. These factors severely limit the utility of log K_D in comparing different micelle-catalyzed reactions.

8.5 Raghavan and Srinivasan's model

P.S. Raghavan and V.S. Srinivasan [82] observed that the cationic micelles of cetyltrimethylammonium bromide (CTAB), cetyldibenzylammonium chloride (CDBAC) and cetylpyridinium chloride (CPC) stabilize the tetrahedral intermediate formed on the hydrolysis of carboxylic esters. Based on this observation a model for bimolecular micellar catalyzed reactions was developed. The model predicts the constancy of k_{obs} values at high detergent concentrations of and can be used to estimate the binding constants of reactants. **Figure 21** shows the distribution of both reactants and nucleophiles in the aqueous and micellar phases proposed as by Raghavan and Srinivasan.

The formations of products are assumed to result from decomposition of ternary complex in micellar phase as well as from the reaction between the substrate and the nucleophile in aqueous medium. After analyzing the data based on this model, they concluded that almost all the nucleophile is present in the bulk phase.

Where D, S and N refer to the detergent monomer, substrate and the nucleophile, respectively, while D_nS and D_nSN are the binary and ternary complexes, respectively. The rate law for the above **Figure 21** is

$$\text{rate} = k_w[S]_0[N]_0 + k_m[D_nSN] \quad (20)$$

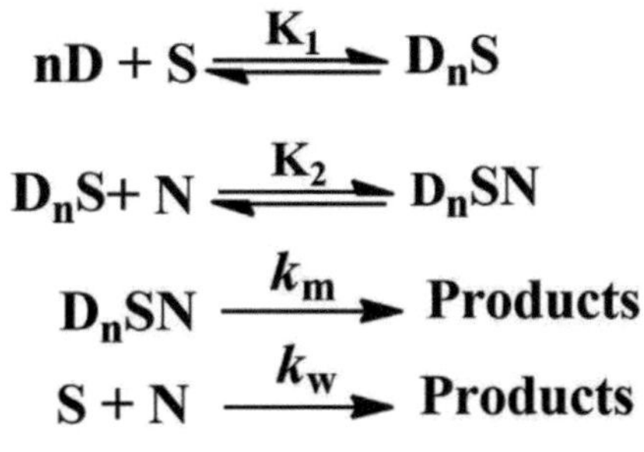

Figure 21.
Reaction steps involved in the Raghavan-Srinivasan's model.

Where, subscript '0'refers to free species. Now,

$$[D_nSN] = K_2[D_nS][N]_0 \text{ and } [D_nS] = K_1[D]^n[S]_0$$

Again, free concentration of substrate,

$$[S]_0 = [S]_T - [D_nS]$$

Where, $[S]_T$ is the total concentration. Thus, the concentration of D_nS and D_nSN can be given as

$$[D_nS] = \frac{K_1[D]^n[S]_T}{1 + K_1[D]^n}$$

$$[D_nSN] = \frac{K_1K_2[D]^n[S]_T[N]_0}{1 + K_1[D]^n}$$

Assuming $[N]_0 = [N]_T - [D_nSN]$. On introducing $[S]_0$, $[N]_0$ and $[D_nSN]$ in Eq. (20), it becomes

$$\text{rate} = \frac{k_w[S]_T[N]_T + k_mK_1K_2[D]^n[S]_T[N]_T}{1 + K_1[D]^n\{1 + K_2[S]_T\}}$$

$$\text{or, } k_{obs} = \frac{k_w + k_mK_1K_2[D]^n}{1 + K_1[D]^n\{1 + K_2[S]_T\}} \tag{21}$$

Eq. (21) may be rearranged in the form

$$\left\{\frac{k_{obs} - k_w}{k_{obs}}\right\}\frac{1}{[D]^n} = K_1K_2\left(\frac{k_m}{k_{obs}}\right) - K_1\{1 + K_2[S]_T\} \tag{22}$$

Eq. (22) predicts a linear relationship between $\{(k_{obs}\text{-}k_w)/k_{obs})\}(1/[D]^n)$ and (k_m/k_{obs}). In this model, the value of the cooperativity index (n) used is taken from the Piszkiewicz's cooperativity model. The values of the binding constants (K_1 and K_2) and K_D (reciprocal of K_1) were calculated from the intercept and slope of the linear plot. It is apparent that the binary complex (D_nS) dissociation constant, K_D in the Piszkiewicz model is actually the reciprocal of K_1 in the Raghavan and Srinivasan model. The K_D values assessed in this model are consistent with those obtained in the Piskiewicz model [4, 83]. However, this model does not provide an explanation of the effect of counter ions or the interaction of products with micelles. When the proposed model is applied to several systems, the value of K_1 and the intercept are approximately equal to $K_2[S]_T \ll 1$ in all cases. The K_1 values evaluated are also consistent with those found in the Piskiewicz model. An abnormal value of K_1 (or, high value of K_1) from both the models would be due to the assumption that the substrate induces micellization in which case there should be very strong association of the substrates with the micelles and thus the equilibrium favors the formation of binary complex or the catalytic micelle, D_nS. Again, the low values of K_2 ($K_2 \ll K_1$) suggest that $[N]_0 = [N]_T$, and therefore, the nucleophile is present almost in full in the bulk phase. Thus, one may suggest that the reaction takes place between the substrate solubilized into a catalytic micelle and the nucleophile residing at the interface.

9. Concluding remarks

This chapter can be divided into two parts: the first part of this chapter addresses the concept of microheterogeneous systems and then various model concepts, their importance and limitations. By microheterogeneous system, one refers to an aggregated system in which the structure of the constituent molecules along with the solvent or other surrounding medium determines the structure of the aggregate. Most of the aggregates are self-assembling and form micelle, reverse micelle, vesicle, liposome etc. Although, each aggregate contains a large number of molecules but the overall aggregate is small enough to form colloidal dimensions and therefore called a microheterogeneous system. These systems have various aspects but the most important feature for the researcher is the study of the effect on reaction kinetics. Surfactants are the most commonly used species that form microheterogeneous systems. The functionality, non-functionality and inertness of a surfactant generally depend on the head group of the surfactant. Most functional surfactants contain nucleophilic, or occasionally basic, residues and form micelles that have a high concentration of reactive groups on their surface. Another very common microheterogeneous system is the reverse micellar system. There are three types of water molecules in the reverse micellar cavities: bound, trapped and free. The water inside the water pool possesses some typically different properties like micro polarity and micro viscosity etc. The rate of the reaction in reverse micelle is generally enhanced as compared to the aqueous medium as well as micelle medium.

It is impossible to carry out micellar chemistry without models and pictures. Thus, this chapter analyses the microheterogeneous systems with the different kinetic models. Several models describe the distribution of solutes and the mutual diffusion of reactants in micellar and reverse micellar media. The distribution of solutes is essentially statistical in nature.

The 'pseudophase model (PP)' is well-treated to explain bimolecular reaction rates. According to this model, in most cases, the rate increase is due to an increase in the concentration of reactive species locally in the micelles (inner or interface region) rather than an increase at the surface. Whereas, in the 'pseudophase ion exchange model (PIE)', adsorption of ionic reactants on the micelle surface occurs and exchange of counter ions is possible which determines the nature of their distribution. Salt also plays an important role in changing the reaction rate. In this case, the micelles are treated as ion exchange resins and competing binding/exchange parameters are used to estimate the rate increases. Paszkiewicz described the positive cooperativity behavior of enzyme-catalyzed reactions in his model and explained the sigmoid shape of reaction rates. Raghavan and Srinivasan proposed the distribution of both reactants and nucleophiles in the aqueous and micellar phases for bimolecular micellar catalysis reactions and products from the micellar and aqueous phases. In most cases, binding constants and dissociation constants in both models, Paczkiewicz and Raghavan–Srinivasa corroborate each other. Although each model has limitations, the models should not have too much trouble interpreting kinetic raw data and to understand the role of models on reaction kinetics requires further studies.

Acknowledgements

I thank Sree Chaitanya College, Habra for giving permission to write the dissertation. In addition, I thank Professor Ambikesh Mahapatra for very useful discussions on all the major points of this chapter. The author acknowledges Sourita Mandal and Soumini Mandal for providing sufficient time for literature survey and chapter writing.

Author details

Harasit Kumar Mandal
Sree Chaitanya College, Habra, West Bengal, India

*Address all correspondence to: harasitju@gmail.com

References

[1] Tascioglu S. Micellar solutions as reaction media. Tetrahedron. 1996;**52**: 11113-11152

[2] Hashimoto S, Thomas JK. Fluorescence study of pyrene and naphthalene in cyclodextrin-amphiphile complex systems. Journal of the American Chemical Society. 1985;**107**: 4655-4662

[3] Mandal HK, Dasmandal S, Mahapatra A. Micellar and mixed-micellar effects on alkaline hydrolysis of *tris*(1,10-phenanthroline)iron(II): kinetic exploration. Journal of Dispersion Science and Technology. 2016;**37**:555-564

[4] Patel BK, Mandal HK, Rudra S, Mahapatra A. Evidence of positive co-operativity in the micellar catalysis electron transfer reaction. Journal of Molecular Liquids. 2018;**250**:103-110

[5] Frendler JH, Frendler EJ. Catalysis Micellar and Macromolecular Systems. New York: Academic Press; 1975

[6] Tang SS, Chang GG. Nucleophilic aromatic substitution of glutathione and 1-chloro-2,4-dinitrobenzene in reverse micelles. A model system to assess the transition-state stabilization in glutathione transferase catalyzed conjugation. The Journal of Organic Chemistry. 1995;**60**:6183-6185

[7] Mandal HK, Majumdar T, Mahapatra A. Kinetics of basic hydrolysis of tris(1,10-phenanthroline) iron(II) in macromolecular assemblies of CTAB. International Journal of Chemical Kinetics. 2011;**43**:579-589

[8] Inoue T, Ebina H, Dong B, Zheng L. Electrical conductivity study on micelle formation of long-chain imidazolium ionic liquids in aqueous solution. Journal of Colloid and Interface Science. 2007; **314**:236-241

[9] Mitra D, Chakraborty I, Bhattacharya SC, Moulik SP. Interfacial and solution properties of tetraalkylammonium bromides and their sodium dodecyl sulfate interacted products: A detailed physicochemical study. Langmuir. 2007;**23**:3049-3061

[10] Fernandez I, Perez-Juste J, Herves P. Cationic mixed micelles as reaction medium for hydrolysis reactions. Journal of Solution Chemistry. 2015;**44**: 1866-1874

[11] Bunton CA, Moffatt JR. A quantitative treatment of micellar effects in moderately concentrated hydroxide ion. Langmuir. 1992;**8**:2130-2134

[12] Vasic-Racki D, Kragl U, Liese A. Benefits of enzyme kinetics modelling. Chemical and Biochemical Engineering Quarterly. 2003;**17**(1):7-18

[13] Rusanov AI. The mass-action-law theory of micellization revisited. Langmuir. 2014;**30**:14443-14451

[14] Posa M. Multiphase separation model for binary mixed micelles. Journal of Molecular Liquids. 2019;**288**:111019

[15] Schramm LL, Stasiuk EN, Marangoni DG. Surfactants and their applications. Annual Report Progress in Chemical Section C. 2003;**99**:3-48

[16] Sekhon BS. Surfactants: Pharmaceutical and medicinal aspects. Journal of Pharmaceutical Technology, Research Management. 2013;**1**:43-68

[17] Corazza M, Lauriola MM, Zappaterra M, Bianchi A, Virgili A.

Surfactants, skin cleansing protagonists. Journal of the European Academy of Dermatology and Venereology. 2010;**24**: 1-6

[18] Blus K, Bemska J. Effect of nonionic surfactants on the dyeing process of cellulose fibres with C.I. Reactive Blue 217. AUTEX Research Journal. 2010;**10**: 64-68

[19] Fendler JH. Membrane Mimetic Chemistry: Chracterizations and Applications of Micelles, Microemulsions, Monolayers, Bilayers, Vesicles and Host-Guest Systems. New York: Wiley; 1983

[20] Hossel P, Dieing R, Norenberg R, Pfau A, Sander R. Conditioning polymers in today's shampoo formulations – efficacy, mechanism and test methods. International Journal of Cosmetic Science. 2000;**22**:1-10

[21] Murawska M, Wiatr M, Nowakowski P, Szutkowski K, Skrzypczak A, Kozak M. The structure and morphology of gold nanoparticles produced in cationic gemini surfactant systems. Radiation Physics and Chemistry. 2013;**93**:160-167

[22] De S, Aswal VK, Goyal PS, Bhattacharya S. Small-angle neutron scattering studies of different mixed micelles composed of dimeric and monomeric cationic surfactants. The Journal of Physical Chemistry B. 1997; **101**:5639-5645

[23] Tung SH, Lee HY, Raghavan SR. A facile route for creating "reverse" vesicles: Insights into "reverse" self-assembly in organic liquids. Journal of the American Chemical Society. 2008; **130**:8813-8817

[24] Wu Z, Yan Y, Huang J. Advanced molecular self-assemblies facilitated by simple molecules. Langmuir. 2014;**30**: 14375-14384

[25] Thakkar K, Bharatiya B, Aswal VK, Bahadur P. Aggregation of 1-alkyl-3-methylimidazolium octylsulphate ionic liquids and their interaction with Triton X-100 micelles. RSC Advances. 2016;**6**: 80585-80594

[26] Galgano PD, El Seoud OA. Micellar properties of surface active ionic liquids: A comparison of 1-hexadecyl-3-methylimidazolium chloride with structurally related cationic surfactants. Journal of Colloid and Interface Science. 2010;**345**:1-11

[27] Rauniyar BS, Bhattarai A. Study of conductivity, contact angle and surface free energy of anionic (SDS, AOT) and cationic (CTAB) surfactants in water and isopropanol mixture. Journal of Molecular Liquids. 2021;**323**:114604

[28] Ruiz CC. A photophysical study of micellization of cetyltrimethylammonium bromide in urea-water binary mixtures. Molecular Physics. 1995;**86**:535-546

[29] Rodriguez A, Munoz M, del Mar Graciani M, Moya ML. Kinetic micellar effects in tetradecyltrimethylammonium bromide–pentanol micellar Solutions. Journal of Colloid and Interface Science. 2002;**248**:455-461

[30] Mandal HK, Patel BK, Dasmandal S, Mahapatra A. Kinetic investigations on the alkaline hydrolysis of *tris*-(1,10-phenenthroline)Fe(II) with guar gum–surfactant interactions. Journal of Dispersion Science and Technology. 2018;**39**:552-559

[31] Rodenas E, Dolcet C, Valiente M, Valeront EC. Physical properties of dodecyltrimethylammonium bromide (DTAB) micelles in aqueous solution and

their behavior as the reaction medium. Langmuir. 1994;**10**:2088-2094

[32] Chatterjee A, Moulik SP, Sanyal SK, Mishra BK, Puri PM. Thermodynamics of micelle formation of ionic surfactants: A critical assessment for sodium dodecyl sulfate, cetyl pyridinium chloride and dioctyl sulfosuccinate (Na Salt) by microcalorimetric, conductometric, and tensiometric measurements. The Journal of Physical Chemistry B. 2001;**105**: 12823-12831

[33] Mukhim T, Dey J, Das S, Ismail K. Aggregation and adsorption behavior of cetylpyridinium chloride in aqueous sodium salicylate and sodium benzoate solutions. Journal of Colloid and Interface Science. 2010;**350**:511-515

[34] Mandal HK, Kundu A, Balti S, Mahapatra A. Kinetic investigation on the oxidation of *tris*(1,10-phenanthroline)iron(II) by oxone: The effect of BSA–SDS interaction. Journal of Colloid and Interface Science. 2012;**378**: 110-117

[35] Pan A, Rakshit S, Sahu S, Bhattacharya SC, Moulik SP. Synergism between anionic double tail and zwitterionic single tail surfactants in the formation of mixed micelles and vesicles, and use of the micelle templates for the synthesis of nano-structured gold particles. Colloids Surface A: Physicochemical Engineering Aspect. 2015;**481**:644-654

[36] Davies DM, Foggo SJ, Paradis PM. Micellar kinetics of acyl transfer from *n*-nonanoyloxybenzenesulfonate and phenyl nonanoate bleach activators to hydrogen peroxide and pernonanoic acid: Effect of charge on the surfactant and activator. Journal of the Chemical Society, Perkin Transactions. 1998;**2**: 1597-1602

[37] Regev O, Zana R. Aggregation behavior of tyloxapol, a nonionic surfactant oligomer, in aqueous solution. Journal of Colloid and Interface Science. 1999;**210**:8-17

[38] Kalyanasundaram K, Thomas JK. Environmental effects on vibronic band intensities in pyrene monomer fluorescence and their application in studies of micellar systems. Journal of the American Chemical Society. 1977;**99**: 2039-2044

[39] Mahmood ME, Al-Koofee DAF. Effect of temperature changes on critical micelle concentration for tween series surfactant. Global Journal of Science Frontier Research Chem. 2013;**13**:1-8

[40] Hait SK, Moulik SP. Determination of critical micelle concentration (CMC) of nonionic surfactants by donor-acceptor interaction with lodine and correlation of CMC with hydrophile-lipophile balance and other parameters of the surfactants. Journal of Surfactants and Detergents. 2001;**4**:303-309

[41] Tiwari S, Mall C, Solanki PP. CMC studies of CTAB, SLS & tween 80 by spectral and conductivity methodology to explore its potential in photogalvanic cell. Surfaces and Interfaces. 2020;**18**: 100427-100434

[42] Dai C, Zhao J, Yan L, Zhao M. Adsorption behavior of cocamidopropyl betaine under conditions of high temperature and high salinity. Journal of Applied Polymer Science. 2014;**131**: 40424-40430

[43] El-Dossoki FI, Abdalla NSY, Gomaa E, Hamza OK. An insight into thermodynamic and association behaviours of cocamidopropyl betaine (CAPB) surfactant in water and water–alcohol mixed media. SN Applied Science. 2020;**2**:690-699

[44] Naskar B, Mondal S, Moulik SP. Amphiphilic activities of anionic sodium cholate (NaC), zwitterionic 3-[(3-cholamidopropyl)dimethylammonio]-1-propanesulfonate (CHAPS) and their mixtures: A comparative study. Colloids and Surfaces B: Biointerfaces. 2013;**112**: 155-164

[45] Acharya DP, Gutierrez JM, Aramaki K, Aratani K-I, Kunieda H. Interfacial properties and foam stability effect of novel gemini-type surfactants in aqueous solutions. Journal of Colloid Interface Science. 2005;**291**:236-243

[46] Cai B, Dong J, Cheng L, Jiang Z, Yang Y, Li X. Adsorption and micellization of gemini surfactants with pyrrolidinium head groups: Effect of the spacer length. Soft Matter. 2013;**9**:7637-7646

[47] Bales BL. A definition of the degree of ionization of a micelle based on its aggregation number. The Journal of Physical Chemistry B. 2001;**105**: 6798-6804

[48] Correa M, Silber JJ, Riter RE, Levinger NE. Nonaqueous polar solvents in reverse micelle systems. Chemical Reviews. 2012;**112**:4569-4602

[49] Buettner CS, Cognigni A, Schröder C, Bica-Schröder K. Surface-active ionic liquids: A review. Journal of Molecular Liquids. 2022;**347**:118160

[50] Nazar MF, Shah SS, Eastoe J, Khan AM, Shah A. Separation and recycling of nanoparticles using cloud point extraction with non-ionic surfactant mixtures. Journal of Colloid and Interface Science. 2011;**363**:490-496

[51] Watanabe H, Tanaka H. A non-ionic surfactant as a new solvent for liquid—liquid extraction of zinc(II) with 1-(2-pyridylazo)-2-naphthol. Talanta. 1978; **25**:585-589

[52] Gu Y, Zhou M, Tu H. Effect of linking groups and hydrophobic groups on properties of sulfate Gemini surfactants. Journal of Molecular Liquids. 2022;**367**:120346-120356

[53] Wong PTT, Weng SF, Mantsch HH. Pressure effect on reversed micelles in water: An infrared spectroscopic study of aqueous dioleoyl phosphatidylethanolamine. Journal of Chemical Physics. 1986;**85**:2315-2319

[54] Persson BO, Drakenberg T, Lindman B. Carbon-13 NMR of micellar solutions. Micellar aggregation number from the concentration dependence of the carbon-13 chemical shifts. The Journal of Physical Chemistry. 1979;**83**: 3011-3015

[55] Nishikido N, Shinozaki M, Sugihara G, Tanaka M. A study on the micelle formation of surfactants in aqueous solutions under high pressure by laser light-scattering technique. II. Journal of Colloid and Interface Science. 1981;**82**:352-361

[56] Liu Q, Ji X, Wang S, Zou W, Li J, Lv D, et al. Effect of additives on surfactant micelle shape transformation: Rheology and molecular dynamics studies. Journal of Physical Chemistry C. 2019;**123**:2922-2932

[57] Sulthana SB, Rao PVC, Bhat SGT, Rakshit AK. Interfacial and thermodynamic properties of SDBS–$C_{12}E_{10}$ mixed micelles in aqueous media: Effect of additives. The Journal of Physical Chemistry B. 1998;**102**: 9653-9660

[58] Scholz N, Behnke T, Genger UR. Determination of the critical micelle concentration of neutral and ionic surfactants with fluorometry, conductometry, and surface tension—

A method comparison. Journal of Fluorescence. 2018;**28**:465-476

[59] Mitsionis AI, Vaimakis TC. Estimation of AOT and SDS CMC in a methanol using conductometry, viscometry and pyrene fluorescence spectroscopy methods. Chemical Physics Letters. 2012;**547**:110-113

[60] Patra N, Ray D, Aswal VK, Ghosh S. Exploring physicochemical interactions of different salts with sodium *N*-dodecanoyl sarcosinate in aqueous solution. ACS Omega. 2018;**3**:9256-9266

[61] Tan CH, Huang ZJ, Huang XG. Rapid determination of surfactant critical micelle concentration in aqueous solutions using fiber-optic refractive index sensing. Analytical Biochemistry. 2010;**401**:144-147

[62] Perinelli DR, Cespi M, Lorusso N, Palmieri GF, Bonacucina G, Blasi P. Surfactant self-assembling and critical micelle concentration: One approach fits all? Langmuir. 2020;**36**:5745-5753

[63] Bunton CA, Nome F, Quina FH, Romsted LS. Ion binding and reactivity at charged aqueous interfaces. Accounts of Chemical Research. 1991;**24**:357-364

[64] Menger FM, Portnoy CE. Chemistry of reactions proceeding inside molecular aggregates. Journal of the American Chemical Society. 1967;**89**:4698-4703

[65] Bunton CA, Fendler EJ, Sepulveda L, Yang K-U. Micellar-catalyzed hydrolysis of nitrophenyl phosphates. Journal of the American Chemical Society. 1968;**90**: 5512-5518

[66] Bunton CA, McAneny M. Catalysis of reactions of p-nitrobenzoyl phosphate by functional and nonfunctional micelles. The Journal of Organic Chemistry. 1977;**42**:475-482

[67] Martinek K, Yataimirski AK, Levashov AV, Berezin IV. In Micellization Solubilization and Microemulsions. In: Mittal KL, editor. Vol. 2. New York: Plenum Press; 1977. p. 489

[68] Berezin IV, Martinek K, Yatsimirskii AK. Physicochemical foundations of micellar catalysis. Russian Chemical Reviews. 1973;**42**:787-802

[69] Laguta AN, Eltsov SV, Mchedlov-Petrossyan NO. Micellar rate effects on the kinetics of nitrophenol violet anion reaction with HO^- ion: Comparing Piszkiewicz's, Berezin's, and Pseudophase Ion-Exchange models. Journal of Molecular Liquids. 2019;**277**: 70-77

[70] Dasmandal S, Mandal HK, Kundu A, Mahapatra A. Kinetic investigations on alkaline fading of malachite green in the presence of micelles and reverse micelles. Journal of Molecular Liquids. 2014;**193**:123-131

[71] Stigter D. On the adsorption of counterions at the surface of detergent micelles[1]. The Journal of Physical Chemistry. 1964;**68**:3603-3611

[72] Bunton CA, Savelli G. Organic reactivity in aqueous micelles and similar assemblies. Advances in Physical Organic Chemistry. 1986;**22**:213-309

[73] Bunton CA. Chapter 13 Reactions in micelles and similar self-organized aggregates. New Comprehensive Biochemistry. 1984;**6**:461-504

[74] Fendler EJ, Fendler JH. Micellar catalysis in organic reactions: Kinetic and mechanistic implications. Advances in Physical Organic Chemistry. 1970;**8**: 271-406

[75] Piszkiewicz D. Micelle catalyzed reactions are models of enzyme

catalyzed reactions which show positive homotropic interactions. Journal of the American Chemical Society. 1976;**98**: 3053-3055

[76] Piszkiewicz D. Cooperativity in bimolecular micelle-catalyzed reactions. Inhibition of catalysis by high concentrations of detergent. Journal of the American Chemical Society. 1977;**99**: 7695-7697

[77] Hill AV. A new mathematical treatment of changes of ionic concentration in muscle and nerve under the action of electric currents, with a theory as to their mode of excitation. The Journal of Physiology. 1910;**40**:190-224

[78] Piszkiewicz D. Positive cooperativity in micelle-catalyzed reactions. Journal of the American Chemical Society. 1977;**99**: 1550-1557

[79] Dunlap RB, Cordes EH. Secondary valence force catalysis. VI. Catalysis of hydrolysis of methyl orthobenzoate by sodium dodecyl sulfate. Journal of the American Chemical Society. 1968;**90**: 4395-4404

[80] Dunlap RB, Ghanim GA, Cordes EH. Secondary valence force catalysis. IX. Catalysis of hydrolysis of para-substituted benzaldehyde diethyl acetals by sodium dodecyl sulfate. The Journal of Physical Chemistry. 1969;**73**: 1898-1901

[81] Bruice TC, Katzhendler J, Fedor LR. Nucleophilic micelles. II. Effect on the rate of solvolysis of neutral, positively, and negatively charged esters of varied chain length when incorporated into nonfunctional and functional micelles of neutral, positive, and negative charge. Journal of the American Chemical Society. 1968;**90**: 1333-1348

[82] Raghavan PS, Srinivasan VS. Kinetic model for micellar catalysed hydrolysis of esters-biomolecular reactions. Proceedings of the Indian Academic Science Chemical Science. 1987;**98**: 199-206

[83] Sen PK, Chatterjee P, Pal B. Evidence of co-operativity in the pre-micellar region in the hydrolytic cleavage of phenyl salicylate in the presence of cationic surfactants of CTAB, TTAB and CPC. Journal of Molecular Catalysis A: Chemical. 2015; **396**:23-30